Diagnostic Radiology Physics
with MATLAB®

Series in Medical Physics and Biomedical Engineering

Series Editors: Kwan-Hoong Ng, E. Russell Ritenour, and Slavik Tabakov

Recent books in the series:

The Physics of CT Dosimetry: CTDI and Beyond

Robert L. Dixon

Advanced Radiation Protection Dosimetry

Shaheen Dewji, Nolan E. Hertel

On-Treatment Verification Imaging: A Study Guide for IGRT

Mike Kirby, Kerrie-Anne Calder

Modelling Radiotherapy Side Effects: Practical Applications for Planning Optimisation

Tiziana Rancati, Claudio Fiorino

Proton Therapy Physics, Second Edition

Harald Paganetti (Ed)

e-Learning in Medical Physics and Engineering: Building Educational Modules with Moodle

Vassilka Tabakova

Diagnostic Radiology Physics with MATLAB®: A Problem-Solving Approach

Johan Helmenkamp, Robert Bujila, Gavin Poludniowski (Eds)

For more information about this series, please visit: https://www.routledge.com/Series-in-Medical-Physics-and-Biomedical-Engineering/book-series/CHMEPHBIOENG

Diagnostic Radiology Physics with MATLAB®

A Problem-Solving Approach

Edited by
Johan Helmenkamp
Robert Bujila
Gavin Poludniowski

CRC Press
Taylor & Francis Group
Boca Raton London New York

CRC Press is an imprint of the
Taylor & Francis Group, an **informa** business

First edition published 2021
by CRC Press
6000 Broken Sound Parkway NW, Suite 300, Boca Raton, FL 33487-2742

and by CRC Press
2 Park Square, Milton Park, Abingdon, Oxon, OX14 4RN

First issued in paperback 2022

Visit the Taylor & Francis Web site at
http://www.taylorandfrancis.com

and the CRC Press Web site at
http://www.crcpress.com

ISBN: 978-0-367-64262-4 (pbk)
ISBN: 978-0-8153-9365-8 (hbk)
ISBN: 978-1-351-18819-7 (ebk)

DOI: 10.1201/9781351188197

Typeset in Computer Modern font
by KnowledgeWorks Global Ltd.

Visit the eResources: www.routledge.com/9780815393658

For our families.

Your support made this book possible.

Contents

CHRISTIANE SARAH BURTON

ARTUR OMAR

MAGNUS DUSTLER

TOMI F. NANO AND IAN A. CUNNINGHAM

SVEN MÅNSSON

Foreword

Diagnostic Radiology Physics with MATLAB®: A Problem-Solving Approach by Johan Helmenkamp, Robert Bujila and Gavin Poludniowski is the second book in the Series in Medical Physics and Biomedical Engineering dedicated to software used by clinical medical physicists.

Given the ongoing and accelerating development of medical device technology and user protocols, fulfilling the mission and delivering the full competence profile of the clinical medical physicist has become a daunting task (Caruana et al. 2014; Guibelalde et al. 2014). However, well-written software and programming skills can help in a multitude of ways (Ferris et al. 2005; Lyra et al. 2011; Donini et al. 2014; Nowik et al. 2015). Regrettably, few suitable texts are available, with the result that the acquisition of high-level programming skills by students and clinical scientists is often a hit-or-miss affair. Few books include exemplary scripts illustrating application to the clinical milieu whilst didactic approaches are insufficiently comprehensive or low in communicative power. *Diagnostic Radiology Physics with MATLAB®: A Problem-Solving Approach* aims to fill this gap with the use of MATLAB® in Diagnostic radiology physics. In this book, the university tutor will find structured teaching text and real-world case study examples ('examples from the trenches') with which to enhance presentations and to set as learning tasks. On the other hand, the student will find a pedagogically appealing and engaging manuscript for individual study whilst the practicing clinical medical physicist will find a learning tool for further development of his or her own skills. The authors have included much practical advice ranging from the importation, manipulation and display of DICOM Data in MATLAB to integration of MATLAB with other programming languages, regulatory issues when deploying software in the clinical environment and the sharing and licensing of software.

Johan Helmenkamp, Robert Bujila and Gavin Poludniowski are busy medical physicists with lots of experience in using MATLAB and I sincerely thank them for finding the time within their busy schedule to dedicate to this important educational initiative and to share their expertise with the readers of this book. I also thank the excellent team of contributors who have presented us with an exciting kaleidoscope of relevant real-world applications. I am sure it has not been easy and I appreciate it. Finally, I would like to wish readers of this text many happy programming hours—and to remind them that programming is power!

Carmel J. Caruana, PhD, FIPEM
Professor and Head, Medical Physics Department, University of Malta
Past Chair, Education and Training Committee, European Federation of Organizations for Medical Physics
Past Associate Editor for Education and Training: *Physica Medica—European Journal of Medical Physics*
Past Member, Accreditation Committee: International Medical Physics Certification Board

Preface

BEFORE stating what kind of book this *is*, we will be clear what it is *not*. This is not a book providing a comprehensive introduction to MATLAB. There are other books for that. Nor is this a book on image processing; you will not find the unsharp mask algorithm explained here. But there are other books for that. And despite the word *physics* in the title, this is also not a book about the physical principles underlying clinical radiology. You guessed it... There are other books for that.

The title is *Diagnostic Radiology Physics with MATLAB®: A Problem-Solving Approach*. In this context, *Diagnostic Radiology* includes diagnostic and interventional radiology based on ionizing sources (x-rays) and non-ionizing sources (Ultrasound and Magnetic Resonance Imaging). Nuclear Medicine is not explicitly addressed, as that is to be covered in another book in the series (though some material in this book is also relevant to that sub-discipline). Although we have aimed the book primarily at medical physicists and engineers and the problems they face, it will undoubtedly also be of interest to technically-minded individuals in other healthcare professions. Some of the material straddles the line between clinical development and clinically-oriented research, so the book will additionally be of interest to students and researchers in university environments.

The book title begs a further question: what do we mean by a *problem-solving approach*? We should point out that we have not provided a large set of problems for the reader to solve themselves. Rather, we have provided examples of how other people have solved *their* problems. This includes a variety of relatively short examples in the first part of the book and in-depth case studies in the second part (eleven of them). Note that the authors of the case studies are not software engineers: the code was written primarily by medical physicists, for medical physicists. The code may not always be implemented in the most efficient way, or developed to software industry standards, but it has the great virtue that it *does the job*. There are benefits from adopting good practices, however, and there are times when software *must* be developed to industry standards. This is also covered.

You may still be wondering... why MATLAB? Well, MATLAB has been around for over 30 years. It is a mature product with a large and ever increasing collection of functionality and toolboxes. The core programming language is simple, powerful and relatively easy to get started with. The MATLAB desktop environment that the user works with is great too. MATLAB is widely used and many have experience of using it from their university studies. However, there is no *special* reason for choosing MATLAB; rather, it is just a good choice. The MATLAB code discussed in this book is all conveniently made available to the user in a software repository (`https://bitbucket.org/DRPWM/code`).

The book is split into two sections: *General topics* and *Problem-solving: examples from the trenches*. The chapters in the first section read well in order, but can also be dipped into independently. Those in the second section are entirely independent and can be read in any order. Find a topic of interest to you and take a look. We hope you learn as much out of reading the book as we did putting it together!

Johan Helmenkamp, Robert Bujila and Gavin Poludniowski, June 2020

Acknowledgements

Editing and co-authoring this book would not have been possible without the support and patience of our families, who have endured more than a few nights and weekends with our absence for the past three years. Thank you.

Further, we wish to convey our sincerest thanks to all contributors. This book would never have reached the level of quality and usefulness without your painstaking efforts in producing the actual content. We are deeply impressed by your expertise and wealth of knowledge in your respective areas and it has been a privilege and a great learning experience to work together with you on this book.

We have enjoyed truly professional support from the team at CRC Press. Kirsten Barr, Rebecca Davies, Francesca McGowan and Georgia Harrison—you have always been there to aid and guide us. And a special thanks to Prof. Carmel J. Caruana for pitching the idea for the book and for inviting us to consider the task of producing it. Thank you all for your efforts and for believing in the idea for this book and for entrusting us to deliver it.

We would like to direct a special thanks to Shashi Kumar, without who we would still be struggling with LATEX formatting and coding. Your skillful service and LATEX knowledge are truly awesome.

We would also like to thank Paul Ganney for generously giving up his time to review a chapter, and Antti Löytynoja and MathWorks for their helpful support throughout the project.

Lastly, we wish to extend our heartfelt gratitude to our colleagues at the Karolinska University Hospital and to our friends for your continuous support and encouragement. A particular mention goes to Robert Vorbau for helping us with reviewing some material and data sets.

Thank you, everybody.

Johan Helmenkamp, Robert Bujila and Gavin Poludniowski

Contributors

Åsa Rydén Ahlgren
Lund University, Skåne University Hospital
Malmö, Sweden

John Albinsson
Lund University
Lund, Sweden

Jonas Andersson
Umeå University, Dicom Port AB
Umeå, Sweden

Robert Bujila
Karolinska University Hospital
Stockholm, Sweden

Christiane Sarah Burton
Boston Children's Hospital
Boston, Massachusetts, USA

Magnus Cinthio
Lund University
Lund, Sweden

Philip S. Cosgriff
United Lincolnshire Hospitals
Lincolnshire, UK

Ian A. Cunningham
Western University
London, Canada

James D'Arcy
Institute of Cancer Research
London, UK

Simon J. Doran
Institute of Cancer Research
London, UK

Magnus Dustler
Lund University, Skåne University Hospital
Malmö, Sweden

Tobias Erlöv
Lund University
Lund, Sweden

Javier Gazzarri
MathWorks, Inc.
Novi, Michigan, USA

Johan Helmenkamp
Karolinska University Hospital
Stockholm, Sweden

Tomas Jansson
Lund University
Lund, Sweden

Tanya Kairn
Queensland University of Technology,
    Brisbane
Queensland, Australia

Piyush Khopkar
MathWorks, Inc.
Natick, Massachusetts, USA

Josef Lundman
Umeå University, Dicom Port AB
Umeå, Sweden

Sven Månsson
Lund University
Malmö, Sweden

Tomi Nano
Western University
London, Canada

Patrik Nowik
Karolinska University Hospital, Karolinska
    Institutet
Stockholm, Sweden

Artur Omar
Karolinska University Hospital, Karolinska
    Institutet
Stockholm, Sweden

Matthew Orton
Institute of Cancer Research
London, UK

Gavin Poludniowski
Karolinska University Hospital, Karolinska
    Institutet
Stockholm, Sweden

Viju Ravichandran
MathWorks, Inc.
Bengaluru, Karnataka, India

Cindy Solomon
MathWorks, Inc.
Natick, Massachusetts, USA

Yanlu Wang
Karolinska University Hospital, Karolinska
    Institutet
Stockholm, Sweden

Matt Whitaker
Image Owl, Inc., Greenwich
New York, USA

Johan Åtting
Sectra AB
Stockholm, Sweden

I

General topics

## CHAPTER 1

The role of programming in healthcare

Johan Helmenkamp and Robert Bujila

Medical Radiation Physics and Nuclear Medicine, Karolinska University Hospital, Stockholm, Sweden

Gavin Poludniowski

Medical Radiation Physics and Nuclear Medicine, Karolinska University Hospital, Stockholm, Sweden
Department of Clinical Science, Intervention and Technology, Karolinska Institutet, Stockholm, Sweden

CONTENTS

YOU CAN CONSIDER this chapter the "ReadMe" file for this book. Just as with a ReadMe file for a software package, you might be tempted to skip over it and plunge straight in, even though you know you shouldn't. However, we invite you to read the following few pages to get you into the right mindset for the rest of this book. This chapter will explore the role that programming could have in your professional career and the role it has in Medical Physics and healthcare as a whole. Further, it covers some of the initiatives that have been started by our professional societies in the last few years that directly or indirectly relate to the need to upgrade training programs for Medical Physics to include relevant programming skills.

1.1 WHAT PROGRAMMING CAN DO FOR YOU

If you are new to programming, you may be daunted by the task of developing the necessary skills and put-off by the sight of impenetrable lines of code. Writing software does not, however, require remarkable intelligence or any special talent. Admittedly, the skills are not acquired in an instant. It requires a little time and effort to nurture your development. But some returns on the time invested can be seen very quickly. Try it. Spend some time going through the next three chapters of this book:

- Chapter 2: MATLAB fundamentals
- Chapter 3: Data sources in medical imaging
- Chapter 4: Importing, manipulating, and displaying DICOM data in MATLAB

By taking just a few hours to work through these chapters, you will see a whole new domain of possibilities open up. Suddenly, what initially may have seemed a complex and even impractical challenge requiring painstaking work for a team of people can become an afternoon programming session for a single individual.

Before we even thought of putting this book together, we were aware of a real need for developing programming skills in the Medical Physics community. In 2017, we designed a one-day course in programming using a case-study approach. We will quote a participant who left comments in the course evaluation. The person commented that (freely translated into English):

> *"The consequence of not including programming in the curriculum of Medical Physics is likely going to lead to the creation of an A- and a B-team. Some medical physicists will succeed in solving their tasks effectively and efficiently, while others will not".*

This is a blunt and provocative statement. In any reasonably sized medical physics department, there is a room for a range of skills and not everybody needs to be an ace programmer. But you, as an individual, can only gain by developing programming skills. It can help you solve your tasks effectively and efficiently and allow you to do things that were not possible before. Do not let statements like the one above discourage you, let it fuel your ambitions instead. You are capable of doing this and since you are reading this book it means that you are going to prove this to yourself quite soon.

You may be wondering, however: why learn MATLAB rather than some other programming language and does this choice limit me? We will not elaborate specifically on the merits of MATLAB in this chapter, but see the very next chapter for more on that. As to whether the choice limits you, this really need not be the case, as is nicely illustrated later on in the book (see *Chapter 6: Integration with other programming languages and environments*).

1.2 WHAT PROGRAMMING CAN DO FOR YOUR CLINIC: CHANGE THE NATURE OF ROUTINE WORK

According to the *Medical Physics 3.0* initiative[1] by the American Association of Medical Physicists (AAPM), excellence in Medical Physics involves applying scientific principles in the healthcare setting. This means that the decisions or advice provided by a medical physicist should be evidence-based and *data-driven* rather than *opinion-driven*[2]. This requires quantitative analysis. A prerequisite for quantification is access to data. Preferably lots of it. In the traditional approach to the practice of medical physics in radiology—typified as *Medical Physics 1.0*—we get our data by walking out into the clinic (often after-hours, as "vampire physicists") and shove our ion chambers at some machine for a couple of hours, scribbling down numbers on a piece of paper or in a spreadsheet. At the end of the session we get a data point. A single data point. Then we let 6 months or a year pass before we test the piece of equipment again. Given enough machines, this can easily translate into a full-time engagement for a team of physicists. And the outcome of this traditional *modus operandi* is maybe one or two data points per year for each machine. Is this really the most cost-effective and smartest way to use physicists in healthcare? Is it really providing maximum value for patients? Imaging techniques are becoming more advanced and complex and there is also a better appreciation of resource management (cost/benefit/waste). For Medical Physics to thrive in this environment, we need to adapt.

With some programming, tasks such as that described above can be semi- or fully automated, take less than 5 minutes per measurement, and get you one data point *per day*[3]. So, programming skills open up the possibility to automate processes, allowing practice to become more effective, efficient, consistent and above all more relevant to the clinical practice. This also allows medical physicists to spend time on improving the quality of service that would have been spent on routine and repetitive work. The impact that automation has on the effectiveness and efficiency of practices in the clinic really cannot be overstated. Yes, it can involve clever and complex programs. But it can also be as simple as a short script to convert data from one format to another. Below are some relevant examples from this book.

Performance measurements on imaging equipment and automated reporting:

- Chapter 5: Creating automated workflows using MATLAB;
- Chapter 11: Automating quality control tests and evaluating ATCM in computed tomography;
- Chapter 17: Automating daily QC for an MRI scanner.

Data collection for radiation dose surveys and improved dosimetry estimations for patients:

- Chapter 12: Parsing and analyzing Radiation Dose Structured Reports;
- Chapter 13: Methods of determining patient size surrogates using CT images;
- Chapter 14: Reconstructing the exposure geometry in x-ray angiography and interventional radiology.

1.3 WHAT PROGRAMMING CAN DO FOR YOUR CLINIC: ENABLE RESEARCH AND INNOVATION

Research and innovation comes in many flavours. It can be on a small or large scale. It can provide a modest incremental improvement or a revolution in practice. It can be supported by external research funding, carried out as part of your job description as a medical physicist, or even done in your spare time. What programming does is make research and innovation possible–or at least provide a wealth of additional possibilities.

With programming, you can extract information from vendor's data files that you could not access otherwise (see *Chapter 20: Importation and visualization of ultrasound data*). With programming, you can introduce innovative algorithms to investigate the feature you are interested in (see *Chapter 19: Estimation of arterial wall movements*). With programming, you can generate synthetic images for virtual clinical trials and save patients from unnecessary exposure (see *Chapter 15: Simulation of anatomical structure in mammography and breast tomosynthesis using Perlin noise*). With programming, you can model the physics of x-rays and optimize radiographic examinations (see *Chapter 16: xrTk: a MATLAB toolkit for x-ray physics calculations*). With programming, you can create data "pipe-lines" to process the enormous data sets typically required for Artificial Intelligence (AI) applications (see *Chapter 18: Image processing at scale by containerizing MATLAB*).

There is much buzz at scientific conferences these days about AI[4] and *Deep Learning* in particular. AI is not a new concept. However, groundbreaking results have been enabled in the recent decade as the necessary hardware and software have become available, along with large data repositories and innovations in algorithms. The field is beginning to make a deep impact in healthcare. The future roles to be played in this by hospital staff, university-based researchers and the vendors remains unclear. Even if medical physicists do not end up playing a prominent role in the development of clinical solutions, it will require a degree of savviness about the techniques not possible without some familiarity with programming.

As this book is focused on developing programming skills, and since the AI revolution in healthcare is in its infancy, we do not focus on AI examples in this book. However, we believe that studying this book will prepare the reader for making that next step.

1.4 WITH GREAT POWER COMES GREAT RESPONSIBILITY

There is a well-known adage, usually attributed to Spider-man's uncle, that, "With great power comes great responsibility". It is true here. Programming gives you great power. And it is our responsibility, along with Hippocrates, to *do no harm*.

Chapters in this book cover good habits (see *Chapter 7: Good programming practices*) and how to responsibly distribute your software (see *Chapter 8: Sharing software*). Innovation and development is often nurtured by free-wheeling experimentation and a sense of "play", but there is an obvious tension when this has the potential to impact the health or privacy of individuals. This book also provides an overview of our legal responsibilities (see *Chapter 9: Regulatory considerations when deploying your software in a clinical environment*) and a case study describing an attempt to conform to industry standards for software development (see *Chapter 10: Applying good software development processes in practice*).

1.5 CONCLUSION

Wielding even elementary programming skills will equip you with a valuable tool that can help you make a positive impact for patients, staff, the public and your business. You don't need to become a professional software engineer. The returns from learning some programming skills are great for the medical physicist or other healthcare professional. And, importantly. . . programming is not as hard as you think! You will make mistakes. But everyone makes mistakes in programming. . . it is part of the process. And do not forget, if you get stuck or need a shortcut to solve your problem. . . Google it! You will be surprised at how many problems similar to those that you face have already been solved by others.

Oh and by the way, you don't need to start from scratch. All the code in this book is available online in a Git repository that we have set up for you.[1]

---

[1]https://bitbucket.org/DRPWM/code

## CHAPTER 2

MATLAB fundamentals

Javier Gazzarri

MathWorks, Inc., Novi, MI, USA

Cindy Solomon

MathWorks, Inc., Natick, MA, USA

CONTENTS

THIS CHAPTER introduces MATLAB® as a language and desktop environment. Later chapters apply concepts described here to applications in medical imaging. We begin with a description of the environment, variables and data types. We show the typical steps for importing, manipulating and visualizing tabular data, and we recommend some best practices for improving code performance.

2.1 INTRODUCTION

2.1.1 What is MATLAB?

MATLAB (the name derives from MATrix LABoratory) is a high-level programming language for technical computing, data analysis, signal and image processing, software development, control design and many other applications. It provides a natural way to express computational mathematics via an intuitive matrix-based language.

MATLAB originated in the 1970s as an academic exercise devised by Professor Cleve Moler at the University of New Mexico to provide his students with a tool for matrix computation that would not require writing FORTRAN code. Prof. Moler spent the 1979–1980 academic year at Stanford University teaching Numerical Analysis, and introduced his first version of MATLAB to his students. Some of these students worked on Control Theory and Signal Processing, and MATLAB turned out to be directly applicable to these disciplines. One of his students was an acquaintance of another engineering graduate student named Jack Little. In 1983, Jack proposed the creation of a commercial product based on MATLAB, envisioning that the evolution of the personal computer would allow its use by the general public. Jack and his colleague Steve Bangert re-wrote the existing MATLAB in C, and the product made its public appearance in 1984 when Little, Moler and Bangert co-founded MathWorks. At present, millions of engineers and scientists worldwide use MATLAB in industry, research and academia.

This chapter describes functionality that is native to MATLAB. In addition to base MATLAB, add-on toolboxes provide advanced application-specific functionality for curve-fitting, statistics, image processing, application deployment, controls, pharmaceutical and financial modeling and many other tasks.

2.1.2 Environment

By default, the MATLAB desktop interface (Figure 2.1) is divided into four main components: the *Current Folder Browser*, the *Command Window*, the *Workspace Browser* and the *MATLAB Editor*. The Current Folder Browser lets you interactively manage files and folders. The Current Folder Browser indicates which files MATLAB can access. The Command Window accepts MATLAB commands and returns the result in the same window. The Workspace Browser shows a list of the variables currently in memory, along with their main features such as size and range.

The MATLAB Editor is the environment typically used to write programs and develop algorithms using a text-based workflow. A script is a sequence of MATLAB commands meant to be reused. It consists of MATLAB statements written sequentially as they would be used at the Command Window. A script can be executed in its entirety using the *Run* button in the toolstrip, or section by section. Sections are defined by double percentile signs (%%) in traditional m-file scripts (.m) or *section breaks* in Live Scripts (.mlx). A plain text script yields results in the Command Window with separate windows for graphs and other types of output, whereas a Live Script combines code, results and documentation in one place.

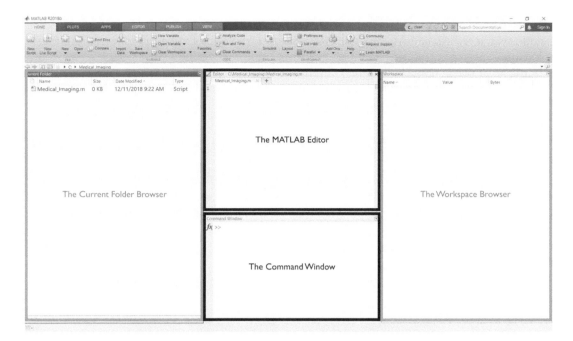

Figure 2.1: The default MATLAB desktop environment.

2.2 VARIABLES AND DATA TYPES

MATLAB stores information in the form of variables with various data types. The default numeric data type in MATLAB is the double-precision floating-point array (referred to simply as a *double array*). Many other data types are available, including tables, structures, cell arrays, character arrays and strings.

MATLAB is a general-purpose language, and consequently, does not require users to predefine a variable data type. Once a value is assigned, MATLAB defines it as the default type compatible with the assigned value. This is known as *dynamic typing* (as opposed to *static typing*).

To initialize a simple variable, use an assignment statement in the form:

$$\texttt{variableName = expression}$$

For example, type the following at the MATLAB prompt ($>>$) in the MATLAB Command Window and press *Enter*:

```
heartRate = 65
```

You should observe output similar to the following and see that the variable `heartRate` in the MATLAB Workspace:

```
heartRate =

    65
```

The output will be shown in the same window, unless it is suppressed by terminating the statement with a semicolon (;). Suppressing the output speeds up program execution. The values of variables (without the variable name) can also be displayed to the window using the MATLAB function `disp`, for example, `disp(heartRate)`.

Variables are case-sensitive and must begin with a letter. You should endeavor to create code that is easily readable by others, for example by naming variables in a meaningful way. Variables are often named in *camelCase* convention, as will be used throughout this book, but several other popular naming conventions exist.

If you enter a computational statement but do not explicitly assign it to a variable, it will be given the default variable name **ans** as shown on the next page:

```
60+5
```

```
  ans =

      65
```

To define a number as a data type different from the default *double*, you can type:

```
singleParam = single(45.678)
```

```
  singleParam =

    single

    45.6780
```

In this case the variable is stored as a floating-point number of lower precision. A *single* consists of 4 bytes rather than the 8 bytes of a double. It is also possible to convert a number from one data type to another:

```
doubleParam = double(singleParam)
```

```
  doubleParam =

    45.6780
```

2.3 ARRAYS AND MATRIX MANIPULATION

All MATLAB variables (even scalar values) are arrays. An array is a collection of values organized into rows and columns and referred to by a single name. To demonstrate, type the command **whos** in the MATLAB Command Window:

```
whos
```

The output displays information about the existing variables in the MATLAB Workspace. Note that the *Size* is 1x1 since these variables are scalars.

Name	Size	Bytes	Class	Attributes
ans	1x1	8	double	
heartRate	1x1	8	double	

To create a row array, define a variable whose elements are separated with a space or comma and enclosed within square brackets:

```
heartRate = [55 57 62 58 60 63 58 58 59 60];
```

All values in an array can be processed with a single function. For example, we can calculate the average heart rate of that array with the **mean** function, which is built into MATLAB:

```
avgHeartRate = mean(heartRate)
```

```
avgHeartRate =

         59
```

A vector describes an array with one dimension, whereas a matrix describes an array with two or more dimensions. To create a matrix with several rows, separate them with semicolons:

```
B = [[1 4 5];[2 3 5];[0 7 9]]
```

```
B =

    1    4    5
    2    3    5
    0    7    9
```

Elements within a matrix can be accessed either by their (row, column) numbers, or by a single number as if the matrix were a one-dimensional array whose length is equal to the total number of elements in it. Because MATLAB is *column-major*, elements are allocated one column at a time. This means that in matrix **B** above, the 4th element has the value 4. Both of the following commands yield the same value:

```
B(4)
B(1,2)
```

```
ans =

       4
```

To perform matrix operations, such as calculating the inverse of a matrix, we simply write:

```
B^-1
```

```
ans =

    0.8000    0.1000   -0.5000
    1.8000   -0.9000   -0.5000
   -1.4000    0.7000    0.5000
```

To treat the matrix as a collection of individual numbers instead and calculate the inverse of each element, we add a period (.) before the operation:

```
B.^-1
```

```
ans =

    1.0000    0.2500    0.2000
    0.5000    0.3333    0.2000
    Inf       0.1429    0.1111
```

Note that an operation returns infinity (**Inf**) if the result is too large to be represented in floating-point format, in this case, $1/0$. Another possible non-numeric result is **NaN** (not a number), which arises in the case of an undefined result, such as $0/0$.

Matrix multiplication is represented by an asterisk (*):

```
B * B

  ans =

      9    51    70
      8    52    70
     14    84   116
```

Conversely, element-by-element multiplication would be:

```
B .* B

  ans =

      1    16    25
      4     9    25
      0    49    81
```

2.4   MORE DATA TYPES

Data does not have to be strictly numeric. For example, text can be stored in a *character array*. These are defined between single quote marks. The sequence `'hello'` is a 1x5 character array, consisting of five letters. Since MATLAB version R2016b, it has also been possible to represent text using a type called *string*. These are defined using double quotes. Each element of a string array contains a sequence of characters. For example, `"hello"` is a 1x1 string array, despite containing five letters. Although it may not be immediately obvious, the introduction of strings has made certain manipulations with text easier.

More abstract types also exist in MATLAB. For example, we can create a *structure*, a data container meant to hold heterogeneous information. We have defined one below with patient information such as name, age, gender and blood pressure. Note that the *fields* of **name** and **gender** have been assigned values using curly brackets (defining a cell array) rather than squares brackets (a numeric array). Numeric arrays can only hold numerical data, whereas *cells* can contain any kind of data type. Here, the elements of the fields of **name** and **gender** are defined between single quote marks, making the content of each cell a character array.

```
patientData.name = {'A','B','C','D','E','F','G','H','I','J'};
patientData.age = [30 43 38 76 79 18 48 44 64 70];
patientData.gender = {'male','male','female','male','male',...
                      'female','female','male','female','female'};
patientData.sysBP = [114 111 132 99 215 193 149 92 109 132];
patientData.diaBP = [64 51 69 51 106 91 73 45 47 66];
```

2.5   CONDITIONAL OPERATORS AND LOGICAL INDEXING

Using a MATLAB feature called logical indexing, we can easily filter information of interest from the original data in `patientData`. For example, if we need to extract all systolic blood pressure values for men, we first turn the gender field into a *categorical* variable so that it can be used as a condition.

```
patientData.gender = categorical(patientData.gender);
```

We can then choose the category `'male'` by using the equality (==) operator. The double equal sign indicates that we are expressing a condition instead of assigning a value. The output of the (==) operator is of type *boolean*, and you can use it as a logical index.

```
maleIndex = patientData.gender == 'male'

  maleIndex =

    1x10 logical array

    1   1   0   1   1   0   0   1   0   0
```

When this logical array is used to index another array, it returns a new array containing only the elements corresponding to the values of one.

```
maleSysBP = patientData.sysBP(maleIndex)

  maleSysBP =

      114   111    99   215    92
```

This allows us to easily calculate the average systolic blood pressure for men:

```
avgMaleSysBP = mean(patientData.sysBP(maleIndex))

  avgMaleSysBP =

    126.2000
```

In addition to ==, there exists a range of other conditional operators in MATLAB. These include: > (greater than), >= (greater or equal to), < (less than), <= (less or equal to) and ~= (not equal to). The results of expressions using these operators are all *boolean* (logical, binary arrays). These logical arrays, themselves, may be used in expressions using *logical operators*, such as the && (and), || (or) or ~ (not). For example, the following returns the logical indexing for male patients with systolic blood-pressure above 140 mm Hg:

```
maleHSysIdx = (patientData.gender == 'male') && (patientData.sysBP > 140)

  maleHSysIdx =

    1x10 logical array

    0   0   0   0   1   0   0   0   0   0
```

Logical indexing is not limited to one-dimensional arrays. An important application in medical imaging is to select features in an image. The parts of an image that are not of interest can be masked-out by assigning a logical index of zero to those pixels. This technique is used both in segmentation and when extracting values from regions-of-interest. An example of segmentation using logical indexing is shown at the end of this chapter.

2.6 CONTROL FLOW

So far, we have shown code snippets where all the code is executed sequentially. In typical computer programming, lines of code may execute in segments based on particular conditions. These traditional programming constructs include *for* and *while* loops, as well as *if-elseif-else* and *switch* statements. In addition, MATLAB has its own unique keywords such as *parfor*, which accelerates computations by distributing them over multiple processor cores. To demonstrate some of these constructs, we will define a 5×5 matrix, **A**, using the function `magic`. This function creates a square matrix in which rows, columns and diagonals add up to the same number.

```
A = magic(5)
```

```
A =

    17    24     1     8    15
    23     5     7    14    16
     4     6    13    20    22
    10    12    19    21     3
    11    18    25     2     9
```

If you need to selectively replace elements satisfying a certain condition (e.g., elements greater than 10) with a zero, a traditional programming language might require two *for* loops, one sweeping rows and the other sweeping columns, an *if* statement establishing the condition and an assignment to perform the replacement in case the condition is true. This is possible in MATLAB too:

```
for i = 1:5
    for j = 1:5
        if A(i,j) > 10
            A(i,j) = 0;
        end
    end
end
```

In MATLAB, however, this task can also be performed in just one line of code:

```
A(A>10)=0
```

```
A =

     0     0     1     8     0
     0     5     7     0     0
     4     6     0     0     0
    10     0     0     0     3
     0     0     0     2     9
```

The example above uses a MATLAB feature called *vectorization*, discussed in detail later in this chapter. Subsequent chapters that elaborate on specific image processing topics will also make extensive use of this feature.

If-elseif-else statements are often used in combination. For example, if we added another field to the `patientData` structure of `heartRate` when exercising, we can do the following:

```
patientData.heartRate = [70 115 220 83 90 180 100 103 50 120];
```

(code continues on next page)

```
for i = 1:length(patientData.heartRate)
    age = patientData.age(i);
    maxHeartRate = 220 - age;
    heartRate = patientData.heartRate(i);

    if(heartRate < (maxHeartRate * 0.5))
        disp(['Patient ' patientData.name{i} ' must increase exercise intensity.'])
    elseif(heartRate < (maxHeartRate * 0.85))
        disp(['Patient ' patientData.name{i} ' has reached target heart rate.'])
    elseif(heartRate < maxHeartRate)
        disp(['Patient ' patientData.name{i} ' has reached high '...
            'exercise intensity.'])
    else
        disp(['Patient ' patientData.name{i} ' has exceeded '...
            'recommended exercise intensity.'])
    end
end
```

The output is as follows:

```
Patient A must increase exercise intensity.
Patient B has reached target heart rate.
Patient C has exceeded recommended exercise intensity.
Patient D has reached target heart rate.
Patient E has reached target heart rate.
Patient F has reached high exercise intensity.
Patient G has reached target heart rate.
Patient H has reached target heart rate.
Patient I must increase exercise intensity.
Patient J has reached target heart rate.
```

You can also use *switch* statements which are similar to *if-elseif-else* statements, but are often used when the condition is a known value rather than an expression. For example:

```
inputVal = input('Enter a field from patientData: ', 's');

switch inputVal
    case {'age' 'heartRate' 'sysBP' 'diaBP'}
        fprintf('Average value for %s is %2.2f\n', inputVal, ...
            mean(patientData.(inputVal)))
    otherwise
        fprintf("No average can be calculated.\n")
end
```

Note that that this snippet requires manual input from the user in the Command Window. Example executions of this code for different inputs yield:

```
Enter a field from patientData: heartRate
Average value for heartRate is 113.10
```

and:

```
Enter a field from patientData: name
No average can be calculated.
```

In addition to introducing the *if-elseif-else* and *switch* statements, the last two code snippets have demonstrated the use of **disp** and **fprintf** to display output to the Command Window. The use of the **input** command to obtain manual input from the user has also be introduced. See the MATLAB documentation for more details on the use of these functions.

2.7   USER-DEFINED FUNCTIONS

2.7.1   Functions

Several functions that are in-built to MATLAB have already been mentioned, such as `mean` and `disp`. MATLAB users can write their own functions too. Functions are pieces of code that accept input arguments and return one or more results, making them efficient and reusable. They can be separate files or written inside a script or another function. For example, a simple function that multiplies two numbers could be written as follows:

```
function out = myProduct(a,b)
    out = a .* b;
end
```

We can call this function after defining `a` and `b`:

```
answer = myProduct(a,b);
```

There is no limit to the number of executable statements a function can store. One advantage of functions over scripts is that they use local variables since they have their own workspace. Consequently, they do not overwrite the MATLAB workspace when executed.

It is helpful to name functions in the form of simple verbs that indicate their purpose, and according to a naming convention, such as camelCase. For example: `calculateBodyMassIndex`.

2.7.2   Anonymous functions

Anonymous functions act like regular functions, but they do not need to be defined or maintained as a separate file or section. They are especially useful when you only need to create a simple function with a single executable statement. The definition of the anonymous function creates a workspace variable of a *function handle* type. Input variable(s) are specified using the "at" (@) symbol. We can create a simple example such as the sigmoid function:

$$\text{sigmoid}(x) = \frac{1}{1 + e^{-x}}$$

This can be defined as an anonymous function in MATLAB as follows:

```
sigmoid = @(x) 1./(1+exp(-x))
```

To evaluate it at a particular value:

```
sigmoid(0)
```

```
ans =
    0.5000
```

Although only one executable statement is allowed per anonymous function, several executable statements can be nested for enhanced functionality. For example, we can define a function of x that performs the following integral of the sigmoid function from $-\infty$ to x:

$$\text{sigmoidIntegral}(x) = \int_{-\infty}^{x} \frac{1}{1 + e^{-t}} dt$$

We can define the `sigmoidIntegral` using two anonymous functions in the following way:

```
sigmoidIntegral = @(x) integral(@(t) 1./(1+exp(-t)),-Inf,x)
```

This can then be evaluated at a particular value:

```
sigmoidIntegral(0)
```

```
  ans =
      0.6931
```

Note that **integral** is one of MATLAB's in-built functions for performing numerical integration of definite integrals.

2.8   DATA ANALYSIS

MATLAB is commonly used for data analysis workflows involving importing, preprocessing and analyzing or visualizing data. If the data is tabular and stored in spreadsheets, MATLAB imports it by default as a *table*. To import a spreadsheet by hand, drag it from the Current Folder pane to the Workspace window or use the *Import Data* button. Clicking *Import Selection* brings up a table matching the spreadsheet name (in this case **data1**) in the Workspace. To help automate the import workflow for future data analysis, MATLAB can generate a function to do this. Figure 2.2 shows a screenshot of the Import Tool.

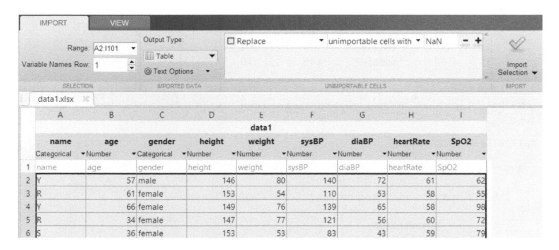

Figure 2.2: The MATLAB Import Tool.

Alternatively, we can import the data programmatically using the MATLAB function **readtable** and assign it the name **patientData** as below:

```
patientData = readtable(fullfile('data','data1.xlsx'));
```

The function **fullfile** creates a location specification for the folder and files in its arguments, inserting platform-specific separators (Microsoft Windows uses backslashes, while Unix-like systems use a forward slash). In this way, the same function can be utilized on all supported platforms to refer to a file location. The data type of the resulting variable **patientData** is a *table*. Like structures, table data can be accessed using dot notation, avoiding the need to remember column numbers. To query the first five rows of the column `age`, enter the following command:

```
patientData.age(1:5)
```

```
    ans =

        57
        61
        66
        34
        36
```

You can also access the table contents with index notation. Below, we display the first three rows of all columns. Because it is a table, MATLAB also displays the header names.

```
patientData(1:3,:)
```

```
    ans =

    3x9 table
        name    age    gender      height    weight    sysBP    diaBP

        'Y'     57     'male'       146        80        140      72
        'R'     61     'female'     153        54        110      53
        'Y'     66     'female'     149        76        139      65
```

MATLAB tables also have properties, such as column names, units, descriptions and custom information that the user can add as needed. To view all properties associated with a table, type:

```
patientData.Properties
```

```
    ans =

      TableProperties with properties:

                  Description: ''
                     UserData: []
               DimensionNames: {'Row'  'Variables'}
                VariableNames: {1x9 cell}
         VariableDescriptions: {}
                VariableUnits: {}
           VariableContinuity: []
                     RowNames: {}
             CustomProperties: No custom properties are set.
        Use addprop and rmprop to modify CustomProperties.
```

2.9 VISUALIZATION

Now that we can import and manipulate data, we can explore different ways to visualize it. For example, if we are looking for correlation between any two variables, we can use the function **plot** to visualize each pair of values on the (x,y) plane. The syntax for the plot function is:

plot(x, y, options)

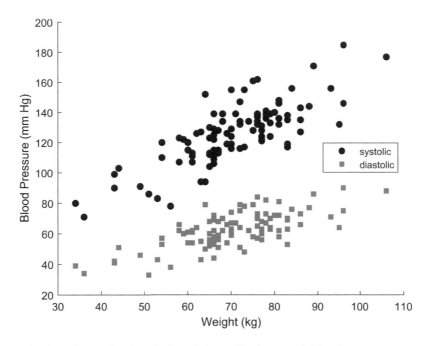

Figure 2.3: A plot of systolic (*circles*) and diastolic (*squares*) blood pressures against weight shows a positive correlation.

We will plot patient systolic and diastolic blood pressures against the patients' weight and observe a positive correlation (Figure 2.3).

```
% Add multiple plots to the same figure
hold on
plot(patientData.weight,patientData.sysBP,'o','MarkerFaceColor','k', ...
    'MarkerEdgeColor','k');
plot(patientData.weight,patientData.diaBP,'s','MarkerFaceColor', ...
    [0.5 0.5 0.5],'MarkerEdgeColor',[0.5 0.5 0.5]);

% Add formatting to the figure
legend({'systolic' 'diastolic'},'Location','best')
xlabel('Weight (kg)')
ylabel('Blood Pressure (mm Hg)')
```

Note the use of formatting options in the code. The `'o'` and `'s'` refer to the type of marker (circle and square) with no connecting lines. The marker colors are specified above with the letter `'k'` for black and the RGB value for gray, `[0.5 0.5 0.5]`. Should we need to repeat this visualization workflow using the same formatting options, MATLAB provides a way to automate the creation of the graph preserving this format. In the figure menu, select *File-Generate Code* to parse through the figure, identify its format and features and create a function. The generated function requires the same type and number of input arguments as the original figure and will output a figure with the same formatting options. Saving the function with the default or a new name will allow us to create graphs of the same style with any other data set. The generated plotting function can be edited and re-saved with new, custom functionality and can also be called from any other function or script, providing a high degree of flexibility. MATLAB offers many other ways to visualize data.

For example, to create a histogram of heart rates, execute the following commands (see Figure 2.4 for the result):

```
histogram(patientData.heartRate, 'FaceColor', [0.5 0.5 0.5], 'LineWidth',1)
xlabel('Resting Heart Rate (bpm)')
ylabel('Patient Count')
```

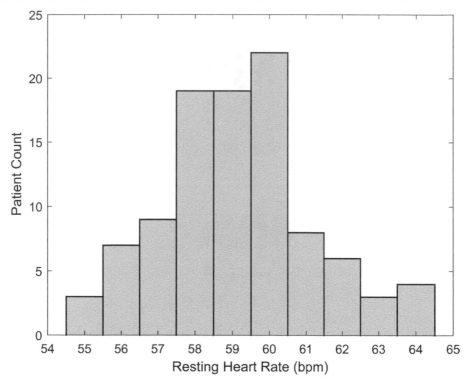

Figure 2.4: A histogram of patient count against resting heart rate.

2.10 HANDLING BIG DATA SETS

Data sets are frequently too large for the available RAM of the computer. MATLAB is able to handle data sets that exceed the computer RAM by means of the *tall* data type. A tall entity (array, table, timetable, cell array, string or cell) is defined using a `datastore`, a descriptive pointer to a location in the computer where the full data set is stored. For example, let us define a tall table T that collects all the spreadsheets in the folder "data".

```
dataStore = datastore('data');
T = tall(dataStore)
```

To minimize computational expense, MATLAB does not initially parse through tall tables. Instead, it defers computation until the user specifies that a result is needed, and at the same time minimizes the number of passes through the data. To demonstrate, use MATLAB to find the mean, standard deviation, minimum and maximum of the systolic blood pressures:

```
meanSysBP = mean(T.sysBP);
stdSysBP = std(T.sysBP);
minSysBP = min(T.sysBP);
maxSysBP = max(T.sysBP);
```

MATLAB queues the calculation until we input the `gather` command:

```
results = gather([meanSysBP, stdSysBP, minSysBP, maxSysBP])

  - Pass 1 of 1: Completed in 9.3 sec
  Evaluation completed in 13 sec

  results =

    124.7220    22.0926    58.0000    195.0000
```

MATLAB calculated four quantities, but only needed one pass through the data.

2.11 CLASSES

MATLAB supports object-oriented programming, a paradigm in which data and functions that operate on that data are encapsulated into objects. Objects are instances of a class (just as Steve is an instance of a patient) and thus share common characteristics in terms of properties (patients have weight and height) and functions performed on them (body mass index depends on weight and height). Functions defined within the class are called *methods* and can operate only on objects of that class. Static class methods are associated with the class but do not need to operate on an object. Object-oriented programming is especially useful for making code more modular and reusable. In the next section, we will define a simple class to demonstrate how Object Oriented Programming works in MATLAB.

2.11.1 Defining a simple MATLAB class

Before creating a class, it is necessary to decide what data and methods all members of that class might have. If we were to define a **Patient** class, all patient objects should have particular properties, such as name, age, height and weight, as well as methods that work on those properties. For example, we might want a helper method **calculateBMI** that uses the values of the height and weight properties to calculate body mass index (BMI).

First, we create a **Patient** class with several properties: **name**, **age**, **height** and **weight**. We want to create objects of this class with initialized values, so we first write a special method called a constructor, which has the same name as the class, to create class objects. We might also want to have a method to calculate the BMI for that patient, assuming the weight and height are in kilograms and meters respectively. Lastly, to print information for a particular patient, we create a method **printPatientInfo** that displays the values of the class properties:

```
classdef Patient
   properties
      name
      age
      height
      weight
   end
   methods
      function obj = Patient(age, name, height, weight) % constructor
         obj.age    = age;
         obj.name   = name;
         obj.height = height;
         obj.weight = weight;
      end
```

(code continues on next page)

```
    function BMI = calculateBMI(obj) % method to return BMI
        BMI = [obj.weight]/[obj.height]^2;
        fprintf("BMI: %2.2f\n", BMI);
    end
    function printPatientInfo(obj) % method to print info
        fprintf("Patient Name: %s\n", obj.name);
        fprintf("Age: %d\n", obj.age);
        fprintf("Height: %d m \n", obj.height);
        fprintf("Weight: %d kg \n\n", obj.weight);
    end
  end
end
```

This class is saved as a separate file called "Patient.m". The following code demonstrates how this class and its methods might be used in a script.

```
% Define an object "Patient X" from class Patient
patientX = Patient(30, "Steve Rogers", 1.88, 99.8);
patientX.printPatientInfo; % Display patient information
patientX.calculateBMI; % Calculate BMI
```

When the code above is executed, it outputs the following in the MATLAB Command Window:

```
Patient Name: Steve Rogers
Age: 30
Height: 1.88 m
Weight: 99.8 kg


BMI: 28.24
```

There is much that could be said about classes and object-oriented programming, however, this introduction is sufficient for the purposes of this book.

2.12   IMPROVING CODE PERFORMANCE

2.12.1   Vectorization

Earlier in this chapter, we explored *vectorization*, a powerful MATLAB capability that speeds up computation and enables compact and efficient coding. Vectorization is one of many coding practices that can increase computation speed. It is in general faster to execute a multiple assignment in vectorized form than by using loops. For example, defining the value of a function using range operators is much more compact and efficient than using a *for* loop. To compare the time it takes in both cases using the *tic-toc* functions for one million points, execute the following lines of code:

```
clear f
tic
for i=1:1e6
    f(i)=exp(-(i-1)/1e6).*sin((i-1)/1e6);
end
toc

clear f
tic
t = 0:1e-6:1;
f = exp(-t).*sin(t);
toc
```

```
Elapsed time is 0.289615 seconds.
Elapsed time is 0.041849 seconds.
```

The vectorized code executes much faster than the original code.

2.12.2 Pre-allocation

When an array is populated within a loop, it is advisable to *pre-allocate* the variable as an array of zeros or NaNs of the expected size of the final matrix. This allocates space in memory for MATLAB to store the newly formed matrix for each step in the iteration, rather than extending the size of the array dynamically. Applying pre-allocation to the previous example results in the following:

```
clear f
tic
f = zeros(1,1e6);
for i=1:1e6
    f(i)=exp(-(i-1)/1e6).*sin((i-1)/1e6);
end
toc
```

```
Elapsed time is 0.056821 seconds.
```

In this instance, the speed-up is significant and the code executes almost as fast as the vectorized operation.

2.12.3 Other best practices

Here are five additional recommendations for improving code readability and execution performance:

- Functions usually run faster than scripts because of the way they use memory and *just-in-time* compilation.
- When a program needs to read or to write from disk, it is best to avoid executing input and output operations in a loop since communication with the hard drive is much slower than with RAM. We recommend saving data as MATLAB variables and saving to disk outside of loops.
- Opening figures within a loop is not advisable since it compromises performance.
- MATLAB reads faster along a column than it does across a row, so sequential data access to a matrix should be performed along columns—think about the loop order when nesting one *for* loop inside another.
- Several MATLAB functions have the option of distributing computation among the computer's multiple cores using parallel computing. This feature is especially useful for accelerating independent calculations.

2.13 EXERCISE—BASIC IMAGE PROCESSING

We will now apply what we have learned so far to do some basic processing on an image. First, we read a *jpg* image from a file and display it:

```
myImage = imread('leaf.jpg');
figure
imshow(myImage)
```

Figure 2.5: Original image.

The result is shown in Figure 2.5. The command we used to display the image was a very useful function called **imshow**. The function we used to import the original *jpg* is **imread**, which stores pixel color level information as a $480 \times 640 \times 3$ array, with the last dimension representing red, green and blue (RGB) levels:

size(myImage)

```
 ans =

     480    640       3
```

In this case, our task will be to replace the background with white pixels and the leaf surface with black pixels, a simple example of image segmentation. How do we know what is leaf and what is not? We will first plot the color levels across the middle of the image, horizontally. We can make three independent plots of the color levels on row 240 (the middle of the image vertically) across all columns using a very compact piece of code:

```
fig = figure;
p4 = plot(1:size(myImage,2), ...
    [myImage(240,:,1); myImage(240,:,2); myImage(240,:,3)]);
p4(1).Color = 'k'; p4(2).Color = 'k'; p4(3).Color = 'k';
p4(1).Marker = 'none'; p4(2).Marker = '.'; p4(3).Marker = 'x';
legend({'R' 'G' 'B'},'Location','best')
ylabel('RGB level across row 240')
xlabel('column')
```

From Figure 2.6 we can see that in all three lines we can clearly distinguish the pixels belonging to the leaf and those belonging to the background. We can now turn the background white and the foreground black, defining background as those pixels with B-component greater than 100 and storing that information as an index. See Figure 2.7 for the result.

```
figure; % Create a new figure
% Define background as those pixels with blue level higher than 100
backgroundIndex = myImage(:,:,3)>100; % generates a (480x480x1) image

% Make background index compatible with image size (480x640x3) using
% repmat, a function that extends the dimensionality of the array specified
% as the first argument (backgroundIndex) along the dimensions indicated
% in subsequent arguments
idx = repmat(backgroundIndex,1,1,3);
myWhiteBackgroundImage = myImage; % Save variable with a different name
myWhiteBackgroundImage(idx) = 255; % Make background white
myWhiteBackgroundImage(~idx) = 0; % Make foreground black
imshow(myWhiteBackgroundImage)
```

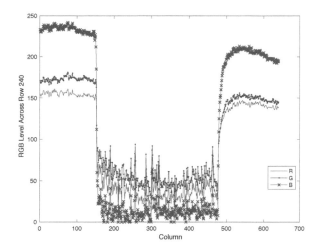

Figure 2.6: Red, green and blue levels for a horizontal line of pixels across the middle line of the figure.

Figure 2.7: Pixels below and above the blue-level threshold of 100 are identified as foreground and background and turned black and white, respectively.

2.14   CONCLUSION

MATLAB is a general-purpose environment for technical computing that provides advanced mathematical functionality without requiring extensive programming experience. This chapter showed the fundamentals of working with MATLAB, including variables, data types, matrix manipulation, statistics and visualization. Subsequent chapters will expand on these key concepts with applications specific to medical imaging.

MATLAB toolboxes used in this chapter:
None

Index of the in-built MATLAB functions used:

categorical	gather	magic	size
clear	histogram	max	std
datastore	hold	mean	tall
disp	imread	min	tic
double	imshow	plot	toc
exp	input	readtable	whos
figure	integral	repmat	xlabel
fprintf	legend	sin	ylabel
fullfile	length	single	zeros

## CHAPTER 3

Data sources in medical imaging

Jonas Andersson and Josef Lundman

Radiation Physics, Department of Radiation Sciences, Umeå University, Umeå, Sweden
Dicom Port AB

Gavin Poludniowski

Medical Radiation Physics and Nuclear Medicine, Karolinska University Hospital, Stockholm, Sweden
Department of Clinical Science, Intervention and Technology, Karolinska Institutet, Stockholm, Sweden

Robert Bujila

Medical Radiation Physics and Nuclear Medicine, Karolinska University Hospital, Stockholm, Sweden

CONTENTS

T HIS CHAPTER serves as an introduction to the data sources that can be used to advance quality assurance and clinical workflows in medical imaging, as well as further research initiatives.

3.1 INTRODUCTION

Medical Physics is a data-driven discipline requiring medical physicists to be knowledgeable about the various sources of data that are available in a clinical setting. Due to the high level of standardization in diagnostic and interventional radiology, many data sources use standards for saving and transferring data, such as Digital Imaging and Communications in Medicine (DICOM)[5]. However, there are also other types of data that may not be presented in such standardized forms, requiring custom solutions to access.

HIS, RIS, PACS and more

Information pertinent to medical imaging can be found from a variety of data sources, consisting of a number of systems that together support the radiological workflow. The main information systems are:

- A *Hospital Information System* (HIS), which is used for high-level hospital administration and is also where electronic records of patients' medical history are stored.

- A *Radiological Information System* (RIS), which is used for patient scheduling and resource management. It is also common that radiologists record their reports in the RIS.

- A *Picture Archiving and Communication System* (PACS), where radiological images are archived. A PACS also provides healthcare professionals access to medical images for review.

- Imaging systems of various modalities where medical images are generated.

- Additional sources such as a *Content Management System* (CMS) containing, for example, an inventory of radiological equipment along with records of maintenance and Quality Control (QC) reports.

HL7 and DICOM

It is common that hospitals procure these various information systems from multiple vendors. For that reason, it is important that standard methods exist to format and handle data, otherwise it would be difficult to integrate systems into the radiological workflow. There are two main relevant standards, namely Health Level 7 (HL7)[6] and Digital Imaging and Communication in Medicine (DICOM)[5].

The HL7 standard provides an interface to communicate alpha-numeric messages between the HIS, RIS and PACS. This allows, for example, communicating of a radiologist's findings back to HIS after review in PACS. However, depending on the vendor, some communication may not adhere to the HL7 standard. HL7 and the interface between HIS, RIS and PACS will not be discussed in further depth here. This aspect of informatics in healthcare is not as central to the role of a medical physicist as that of DICOM.

The DICOM standard was born out of necessity. The lack of a standard for Computed Tomography (CT) and Magnetic Resonance Imaging (MRI) devices in the 1980s lead to practical difficulties in decoding the information in image files. This meant that it was essentially only the manufacturer of the imaging system that could decode and print the acquired images. It also meant that integrating scanners into the clinical workflow was difficult, especially if the clinic had multiple systems from different vendors. To overcome this obstacle, the American College of Radiology (ACR) and National Electrical Manufacturers Association (NEMA) formed a committee in 1983. The result was a standard for formatting, communicating and managing medical image data, published in ACR-NEMA Standards Publication No. 300 (1985). This was a precursor to the present-day DICOM standard. The DICOM standard is still continuing to evolve to conform to current innovations and methods of quality assurance in radiology.

Other kinds of information, such as regarding radiology equipment installation and service history, may be stored in a CMS. This can vary in content and descriptions since there are currently no common standards for this type of information within healthcare.

DICOM metadata and RDSR objects

The most common data that a medical physicist is likely to come into contact with is data generated on an imaging modality. The data will typically conform to the DICOM standard. DICOM provides rules on how to format images and other relevant pieces of information (referred to as *metadata*) associated with an imaging study. Furthermore, it specifies methods for handling the data generated on the imaging system (including networking), and other methods to support workflow such as special DICOM objects called Modality Worklist (MWL), which the RIS can use to send scheduling information to modalities.

The DICOM standard also provides definitions for files that do not contain image information, but instead only provide summary data. For example, the radiation dose indices and associated technique parameters associated with a study are stored in the Radiation Dose Structured Report (RDSR). To a medical physicist, RDSR objects are of great interest, as this information can be used to track or monitor the radiation dose indices in a clinic. An organization promoting health care information system integration, Integrating the Healthcare Enterprise (IHE), provides a technical framework that can be used to set up systems to aggregate RDSR through the Radiation Exposure Monitoring (REM) profile[7].

REM-systems are designed to simplify the collection and communication of information pertinent to the estimation of patient radiation dose indices associated with x-ray imaging procedures. The estimation of radiation dose indices is often a core task for medical physicists, given the legislative requirements in some jurisdictions on assessing radiation dose against Diagnostic Reference Levels (DRLs)[8].

Medical images

Medical physicists can also find important information within the actual images created by radiology equipment. This could be, for example, through imaging phantoms to look for image artefacts or to determine technical metrics such as pixel noise, or by performing calculations on patient imaging data to estimate metrics such as the Size-Specific Dose Estimate (SSDE) used in CT imaging[9, 10]. Analysis of images may naturally also be used in radiology for evaluation of image quality, for dose optimization purposes, both in phantoms and actual patients. Medical images will typically be available in DICOM format, but other formats are possible.

A smorgasbord of other possibilities

Aside from DICOM and RDSR objects, patient-specific information can also be obtained in other ways, for example, by asking each patient and recording their height and weight at the radiology department. If this is recorded in the HIS or RIS, it can later be obtained by querying the HIS/RIS directly and associating those values with the radiation dose reported by modalities.

For quality assurance of radiology equipment, medical physicists can determine the required information in various ways. The conventional method for quality assurance is the manual measurement of physical metrics such as air Kerma and half-value-layer, or indirect metrics via quality assurance instruments, for example, tube potential or total filtration, or image quality metrics. Another source of quality assurance information may be found in the CMS, where reports on installation, preventive and corrective maintenance test results from medical engineers are stored. Yet another option is to access information directly from imaging equipment, where key performance indicators on equipment functionality are collected for immediate display, for example, as is done in autoQA™ for GE Healthcare CT scanners[11], or Siemens Healthineers teamplay[12] and Guardian[13]. In these last examples of non-image data, the file format for saved information may vary widely. In this book, many examples are provided, demonstrating the handling of a wide variety of data sources, including but not limited to, DICOM objects.

3.2 THE DICOM STANDARD AND FILE FORMAT

3.2.1 The DICOM model of the real world

In order to understand the DICOM standard, DICOM's model of the real world needs to be understood. In this model, real-world objects, such as patients and imaging equipment, and their relationship with each other is described. The model places patients at the top of the relationship with other objects. The patient visits, for example, a caregiver, and one or multiple studies are performed on the patient. Each study consists of one or more series, which in turn produces one or more instances. An instance in this contexts could be x-ray images, pathology slices, waveforms, etc. These instances are saved as files. For example, when a patient has an x-ray exam performed, the x-ray machine will produce DICOM files containing the images, and, possibly, also dose reports. See Figure 3.1 for an illustration of the DICOM view of the world.

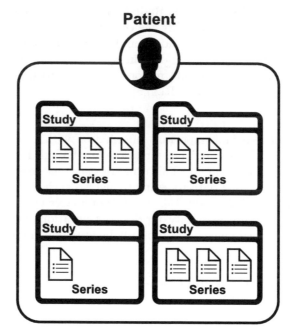

Figure 3.1: Illustration of the DICOM model of the real world which places patients at the top of the relationships with other real-world objects. A patient can have one or multiple studies performed, here illustrated by folders. Each study consist of one or more series, illustrated by a paper. The series, in turn, consists of one or multiple instances, illustrated as bullet lists.

3.2.2   Information object definitions, modules and attributes

A DICOM file is often referred to as an object. DICOM objects are defined by Information Object Definitions (IOD). An IOD is, in essence, a recipe for which pieces of information are relevant to include in the object[1]. A logical grouping of individual pieces of information is called a module. Some of these modules, for example, *Patient*, *General study* and *General series*, are mandatory for most modalities while others may be conditionally required or optional. A module, in turn, comprises a number of individual pieces of information that logically can be grouped together. These individual pieces of information are called attributes. An attribute is identified with a tag, which has a group and element number. For example, the attribute *PatientID* has the group number 0010 and element number 0020. The attribute's group and element are often combined into the following form: (0010,0020). Each attribute essentially represents a single data element in the definition of the DICOM object. When formatted into a file, the data elements come after one another, like a train. Each data element has four parts:

1. The attribute tag.
2. The Value Multiplicity (VM) or the number of data points in an attribute.
3. The Value Representation (VR) or the format of the data in the data field.
4. The data field where the actual data in the attribute resides.

Value representation can be implied, that is, the VR of the attribute does not have to be explicitly given in the data element if the attribute is specified in the DICOM standard. DICOM specifies 28 VRs which cover different ways of formatting text and numeric values. This, for example, includes VRs called "Person Name" and "Date Time". Depending on the VR, attributes can be imported as integers (e.g., uint16), doubles, structs, cells or strings.

Through the IOD, the DICOM standard specifies Type 1, 2 and 3 attributes. A Type 1 attribute entails that the attribute must be present in the DICOM object and the data field needs to be filled out with information. An example of a Type 1 attribute is *PatiendID* (0010,0020). A Type 2 attribute is mandatory to include in the object, however, the data field does not have to be filled out. An example of a Type 2 attribute is *Patient name* (0010,0010). A Type 3 attribute is optional and vendors can choose to include a Type 3 attribute or not. *Study Description* (0008,1030) is an example of a Type 3 attribute. Additionally, DICOM specifies conditional Type 1 or Type 2 attributes (denoted 1C and 2C). These attributes must be specified if some condition is met (e.g., if contrast enhancement is used).

The DICOM standard provides methods for vendors to include non-standard attributes in the DICOM objects they generate. These non-standard attributes are called "Private" attributes. The private attributes that are visible will depend on the model and vendor of the imaging equipment.

Note that DICOM makes use of Unique Identifiers (UIDs) to identify such things as specific instances, studies and series. These UIDs are assigned as the data field values for such attributes. UIDs provide the possibility to uniquely identify an item in the DICOM universe, across countries, sites and vendors.

---

[1]An overview the IODs can be found in Table A.1-1a in DICOM PS3.3: Information Object Definitions: http://dicom.nema.org/medical/dicom/current/output/html/part03.html.

3.2.3 About DICOM images

The image data in a DICOM object is generally stored in the *Pixel Data* tag (7FE0,0010). For DICOM images, it is mandatory to provide a frame of reference, or coordinate system. This frame of reference, allows images from different acquisitions to be related. If you have two or more images with the same *Frame of Reference* UID (tag (0020,0052)), the images will have an identical spatial coordinate system. That is, if the same coordinate values appear in two of these images, they will represent the same point in space. However, it is important to note that time is not part of this coordinate system and movements between acquisitions must be considered.

When there are multiple images in a series, for example, in a CT acquisition, the series may be either divided into multiple files, one for each slice, or all slices may be placed into a single DICOM file. The latter is not common when handling medical x-ray images but more prevalent for MRI. When there are multiple files, most of the metadata will be identical for the different images of the series. However, information such as the (0020,0032) *Image Position (Patient)* will differ as the different slices represent different parts of the patient.

While DICOM is often associated with images, such as CT or MRI scans, other forms of data objects are defined as well. One data object relevant to x-ray imaging is the Radiation Dose Structured Report (RDSR) which summarizes the dose indices and technique parameters associated with an imaging study. More information about its structure and contents can be found in a later chapter (*Chapter 12: Parsing and analyzing Radiation Dose Structured Reports*).

3.2.4 DICOM and MATLAB

DICOM metadata (attributes other than *Pixel Data*) can be imported into MATLAB using the **dicominfo()** function. The *Pixel Data* in a DICOM object can be accessed using the MATLAB function **dicomread()**.

If a variable **dcmInfo** is a *struct* containing the imported image metadata, accessing arbitrary attributes, such as tube voltage, can be done using dot notation, for example: **dcmInfo.KVP**. The function **dicomlookup** can also be used if you know the group and element number of the attribute of interest. For example, **dcmInfo.(dicomlookup('0018', '0060'))** can be used to access the kVp attribute. Attributes with odd group number, for example, 0053, are always private attributes. Private attributes show up in MATLAB as **Private_<Group>_<Element>**.

When importing DICOM objects with MATLAB, special attention should be given to which data types the different attributes are imported as. In many cases, the attributes may be imported as integers and should be re-cast to double, if you are going to do any calculations with those values. This is important to keep in mind as the result of performing calculations with integers will always result in an integer. This is illustrated in the example below, where the metadata for a DICOM image supplied with MATLAB is imported and the number of rows is extracted. This is then used to calculate the midpoint of the first pixel in a row, with respect to the middle of the image.

```
dcmInfo = dicominfo('CT-MONO2-16-ankle.dcm');
rows = dcmInfo.Rows
midOfFirstPixel = -rows/2 + 0.5
```

This provides the output:

```
rows =

  uint16

   512

midOfFirstPixel =

  uint16

   1
```

Something has gone wrong above: the result has been rounded to the nearest positive integer. In fact, it is the type of the variable **rows** that is the problem. It can be identified from the output as uint16 (unsigned integer). If the variable **rows** is first cast to a *double*, then the answer is not rounded. For example:

```
rows = double(dcmInfo.Rows)
midOfFirstPixel = -rows/2 + 0.5
```

This provides the correct output of:

```
rows =

   512

midOfFirstPixel =

  -255.5000
```

Chapter 4 (*Importing, manipulating and displaying DICOM data in MATLAB*) provides more information on MATLAB's in-built functionality for handling DICOM data. Two other chapters (Chapters 6 and 18) demonstrate further possibilities to work with DICOM data, when the native MATLAB functionality proves insufficient or inconvenient.

3.2.5   DICOM communication and data access

In addition to specifying how data should be stored in files, the DICOM standard also specifies a network communications protocol. This protocol uses TCP/IP (i.e., the internet protocol suite) for communication between systems. Here we will only look at a simplified example of how DICOM data is communicated between systems.

In DICOM communication, in addition to the IP-address and port, an application entity title (AET) is used. This is a name that identifies an application entity, that is, a system that uses or provides DICOM services. The name needs to be locally unique, but the same name can be used at different locations.

When, for example, DICOM files are to be sent from an x-ray machine to a PACS, the machine sending the images is called a service class user (SCU) and the PACS receiving the images is called a service class provider (SCP). First, an association is established between the SCU and SCP. In establishing the association, the SCU calls the SCP using the IP-address, port and AET. Then there is a negotiation where the SCP has to allow the SCU to send the files. The SCP might have specific requirements for which types of DICOM files or image compression techniques it receives. If the negotiation is successful, the SCU will send the files to the SCP before sending a release request and releasing the association.

3.2.6 DICOM conformance

Nobody can force a vendor to use the DICOM standard. Further, there is no organization that enforces that a vendor is using the DICOM standard correctly. However, manufacturers of x-ray equipment provide DICOM conformance statements that specify the vendor's particular implementation of the DICOM standard in their product. The conformance statement provides information on which IODs and services are supported on the product. Further, if any private attributes are used, they are specified in the DICOM conformance statement.

Since private attributes are vendor-specific additions, MATLAB will typically need some help in interpreting them. If the user knows the DICOM VR type, the data can be cast as the equivalent MATLAB type using the **typecast** function. One example is the *Noise Index*, which is an important parameter on GE Healthcare CT systems and influences the Automatic Tube Current Modulation (ATCM). Since the Noise Index is vendor specific, DICOM does not have explicit support for it. However, it is included as a private attribute. The DICOM conformance statement can be found on the internet or you can ask your local vendor representative. You can then look up which private attribute corresponds to the Noise Index. At the time of writing, attribute (0027,101F) corresponds to *10*Noise Index*. That is, the Noise Index can be found by accessing the attribute (0027,101F) and dividing by 10. The VR type is SL (Signed Long) which corresponds to the *int32* MATLAB type. Therefore we can extract the noise index by:

```
dcmInfo = dicominfo('exampleImage.dcm');
GE_NoiseIndex = double(typecast(dcmInfo.Private_0027_101f, 'int32'))/10;
```

The file "exampleImage.dcm" is supplied with the code to this chapter. The conversion from *int32* to *double* was necessary to prevent a loss of precision in the division operation. Some other VR types that relate to MATLAB types are: US (*uint16*), SS (*int16*), UL (*uint32*), FS (*single*) and FD (*double*).

Vendors may also interpret the standard in different ways. One such example is how the table position is specified for interventional radiology equipment. A positive value for the table height might mean the table is above or below a default position depending on the manufacturer. The default position may also vary. The DICOM conformance statement may contain information on how the different parameters are defined. In other cases, clarifications must be requested from the manufacturer. It may also be the case that optional tags are included in the DICOM files but placed in private tags instead of in the tag defined in the DICOM standard. The DICOM conformance statement is not always easy to read and will look different between different manufacturers. Ideally, a validation of the DICOM conformance is carried out as a part of commissioning of x-ray equipment. From a quality management perspective, such validation is also valuable to perform continuously, for example, following a software upgrades to imaging equipment.

3.3 OTHER DATA SOURCES

For those working with image data in a clinical environment, handling DICOM data can be a daily occurrence. The reader will, however, encounter other data sources. A brief survey of other formats is given below, along with some of the relevant MATLAB functions to handle such data. For further information, as always, explore the relevant MATLAB documentation.

3.3.1 Other standard image file formats in medical imaging

DICOM is not the only format for medical imaging data. A few formats that MATLAB has the functionality to handle will be discussed below. Some formats, such as DICOM and *Interfile*, were developed to standardize images produced by diagnostic modalities. Others, such as *Analyze* and *Nifti*, emerged to fulfil the needs of post-processing analysis.

Interfile

The Interfile format was introduced in the late 1980s as a vendor-independent way to transfer Nuclear Medicine data between computer systems. It is gradually being replaced by DICOM in clinical use, both for raw and reconstructed data. Typically (as of Interfile V3.3), the voxel data is stored in a binary file with the suffix *.img* and the metadata in a human-readable text header file with suffix *.hdr*. MATLAB provides functions to read Interfile data: `interfileread()` to read in the voxel data and `interfileinfo()` to read the metadata.

Analyze7.5

The Analyze7.5 format was developed at the end of the 1980s at the Mayo Clinic (Rochester, MN, USA). It has seen use for PET and MR modalities, but has largely been superseded by other formats, such a Nifti. An image consists of two files: one with a suffix *.img* and one with a suffix *.hdr*. The image voxel-data is stored in the former, while metadata is stored in the latter. Both are binary files and not human-readable. MATLAB provides functions to read Analyze7.5 data: `analyze75read()` to read in the voxel data and `analyze75info()` to read the metadata.

Nifti

Nifti (Neuroimaging Informatics Technology Initiative) format was developed by the National Institutes of Health (Bethesda, Maryland, USA). It was designed to address weaknesses of the Analyze format, such as the specification of patient orientation. The Nifti format allows data to be stored in two parts (*.img* and *.hdr*), but more typically it is saved as a single *.nii* file. MATLAB provides functions to read and write Nifti data. The functions `niftiread()` and `niftiinfo()` can be used to read in the voxel and metadata, respectively. The `niftiwrite()` function is used to write a new file.

3.3.2 Additional formats you might encounter for image data

While operated in clinical mode, most imaging equipment designed for healthcare environments produce image data in DICOM format. However, not infrequently, we find that when operated in a "service" or "research" mode, a device saves data in a different or even custom format. A device not designed specifically for a clinical environment may also use a non-DICOM format. Finally, even if you have DICOM data, a software package that you intend to use for processing may require conversion to a different file format. Data may be stored in standard (but non-medical) image formats such as JPEG, TIFF and BMP. MATLAB can process images in a wide variety of such image formats, using the `imread()` and `imwrite()` functions.

Image data in custom formats may be human-readable (a text file) or not (a binary or "raw" file). In either case, MATLAB provides the ability to process the files using the `fread()` and `fwrite()` functions. Before using these functions, it is important to know something about your file format, for example, header length and data types. It is preferable to have a full specification of the custom format, although often a partial specification is sufficient, combined with some detective work.

3.3.3 Other types of data sources

Images are not the only data sources of interest to those working in radiology. Tabulated data stored in .txt (plain text) files, .csv (comma-separated values) files or various spreadsheet formats, can be read into or saved from MATLAB as:

- *tables* using the `readtable()` and `writetable()` functions
- *cell arrays* using the `readcell()` and `writecell()` functions
- *arrays* using the `readmatrix()` and `writematrix()` functions

It may also be of interest to extract data from or query databases. MATLAB provides the *Database Toolbox* and functions such as `database()` for connecting to databases. See the MATLAB documentation for more information. A later chapter provides a simple example of using databases (see *Chapter 5: Creating automated workflows using MATLAB*).

3.3.4 Storing data for re-use in MATLAB

Any data in the MATLAB Workspace, such as an array storing an image, can be saved in MATLAB's own *.mat* format using the `save()` function. This is useful for saving data to be used again in another MATLAB session. The data can be loaded again using the `load()` function. The *.mat* data format is portable, so it can be passed to another system or user, as long as the data will be accessed from an instance of MATLAB.

3.4 CONCLUSION

Medical Physics is coming to depend ever more on digital information in various applications such as advanced patient-specific dosimetry[14], general quality assurance and optimization and the collection and benchmarking of Diagnostic Reference Levels (DRLs). For work with quality assurance of radiation dose and image quality, for ensuring legislative demands on radiation safety are met and for research in radiology, the medical physicist should be aware of all the standards, subsystems and information sources discussed here. This includes the connections to other systems in healthcare, containing pertinent information, such as CMS, HIS, RIS and PACS. For research and development, the medical physicist must be familiar both with the data sources and the means to collect and analyze the information.

MATLAB toolboxes used in this chapter:			
Image Processing Toolbox			
Index of the in-built MATLAB functions used:			
analyze75info	fread	load	readtable
analyze75read	fwrite	niftiinfo	save
dicominfo	imread	niftiread	typecast
dicomlookup	imwrite	niftwrite	writecell
dicomread	interfileinfo	readcell	writematrix
double	interfileread	readmatrix	writetable

Importing, manipulating and displaying DICOM data in MATLAB

Piyush Khopkar

MathWorks, Inc., Natick, MA, USA

Josef Lundman

Radiation Physics, Department of Radiation Sciences, Umeå University, Umeå, Sweden
Dicom Port AB

Viju Ravichandran

MathWorks, Inc., Bengaluru, Karnataka, India

CONTENTS

THIS chapter gives a brief overview on the capabilities that MATLAB provides for analyzing medical imaging data. Since the DICOM standard is the established framework for storing and transmitting medical images, this chapter will focus on the native DICOM support in MATLAB.

4.1 INTRODUCTION

MATLAB and the *Image Processing Toolbox* provides an easy way to access and analyze DICOM data. This chapter discusses working with DICOM images in MATLAB, with primary focus on: importing, writing and displaying DICOM data, ways to anonymize data in MATLAB, extracting meta data and techniques on rendering images and volumes.

Each subsection will have sample code for you to try as you read. As a prerequisite for this chapter, it is recommended to familiarize yourself with the fundamentals of MATLAB (Chapter 2) and the DICOM standard (Chapter 3). Some familiarity with basic image processing terminologies will also be assumed.

4.1.1 Locating DICOM data

MATLAB's *Image Processing Toolbox* includes sample DICOM data sets. These data sets cover examples of magnetic resonance imaging (MRI), computed tomography (CT) and ultrasound (US) modalities. There are also a few sample data sets contributed by the MATLAB community on *MATLAB Central* as well, such as: *brain_017.dcm*[1]. For the examples in this chapter, we will mostly use the sample DICOM data sets included with MATLAB and the *Image Processing Toolbox*.

Before importing DICOM data into MATLAB, we will first create a variable **dataDir** and assign the path of our sample DICOM images to this variable:

```
% Assign DICOMDIR variable to the sample DICOM images in MATLAB
dataDir = fullfile(matlabroot,'toolbox/images/imdata');
```

If we want to add **dataDir** to the MATLAB path we can do so as follows:

```
addpath(dataDir);
```

This means that we will not have to provide the full path to files in this location in order for MATLAB to find them. However, this location happens to be on the default MATLAB path already and so this step is unnecessary here. MATLAB's `isdicom` function can be used to check whether a particular data file is a valid DICOM file. The function returns the logical value 1 if the file is a DICOM file or 0 if it is not. There are several examples of the use of `isdicom` throughout this book.

4.1.2 Interpreting DICOM data

The DICOM file format has a provision to store metadata which provides useful information about the image including the data associated with the image capture, the modality and physical attributes of the images such as the size and dimensionality of the image. In MATLAB, the metadata contained in a DICOM file can be inspected using the `dicomCollection`, `dicomdisp` and `dicominfo` functions.

The `dicomCollection` function gathers information on the DICOM files and returns summary metadata as a *table*. Related series of data are aggregated using the value of the `SeriesInstanceUID` data field. The following code snippet shows the syntax for generating the details about the DICOM data at the location **dataDir**:

```
collection = dicomCollection(dataDir)
```

---

[1]This data set can be downloaded from: https://www.mathworks.com/MATLABcentral/fileexchange/ 2762-dicom-example-files. Alternatively, sample DICOM data sets can also be downloaded from www.dicomlibrary.com

The output of the command is shown below:

```
collection =

  5x14 table
```

	StudyDateTime	SeriesDateTime	PatientName	PatientSex
s1	30-Apr-1993 11:27:24	[30-Apr-1993 11:27:24]	``Anonymized''	``''
s2	14-Dec-2013 15:47:31	[14-Dec-2013 15:54:33]	``GORBERG MITZI''	``F''
s3	03-Oct-2011 19:18:11	[03-Oct-2011 18:59:02]	``''	``M''
s4	03-Oct-2011 19:18:11	[03-Oct-2011 19:05:04]	``''	``M''
s5	30-Jan-1994 11:25:01	[]	``Anonymized''	``''

Only four of the fourteen columns are displayed here, due to space considerations. Other useful columns include *Rows*, *Columns*, *Frames*, *Modality* and *Series Description*.

At times, it is useful to be able to inspect the complete metadata for a particular DICOM file. The `dicomdisp` function in MATLAB can be used for this purpose; it reads and displays the metadata from a compliant DICOM file to the MATLAB output window. It will display the following information about each DICOM attribute: *Name*, *Location*, *Level*, *Tag*, *Value Representation (VR)*, *Size* and *Data*. Example:

```
% Display metadata from 'CT-MONO2-16-ankle.dcm'
dicomdisp('CT-MONO2-16-ankle.dcm');
```

This gives the output:

```
File: /Applications/MATLAB_R2019b.app/toolbox/
images/imdata/CT-MONO2-16-ankle.dcm (525436 bytes)
Read on an IEEE little-endian machine.
File begins with group 0002 metadata at byte 132.
Transfer syntax: 1.2.840.10008.1.2 (Implicit VR Little Endian).
DICOM Information object: 1.2.840.10008.5.1.4.1.1.7
(Secondary Capture Image Storage).
```

Location	Level	Tag	VR	Size	Name	Data
0000132	0	(0002,0000)	UL	4 bytes -	FileMetaInformationGroupLength	*Binary*
0000144	0	(0002,0001)	OB	2 bytes -	FileMetaInformationVersion	*Binary*

For purposes of brevity we do not produce the entire output of `dicomdisp` and have truncated it after the first two attributes.

When the metadata needs to be extracted programmatically, in a script or function, the `dicominfo` function can be used. This function returns a *structure* with the complete DICOM metadata. To read the metadata from "CT-MONO2-16-ankle.dcm" into a variable called `dcmInfo`:

```
% Read metadata from 'CT-MONO2-16-ankle.dcm'
dcmInfo = dicominfo('CT-MONO2-16-ankle.dcm')
```

This will display:

```
dcmInfo =

  struct with fields:

            Filename: '/Applications/MATLAB_R2019b.app/toolbox/
                       images/imdata/CT-MONO2-16-ankle.dcm'
         FileModDate: '18-Dec-2000 12:06:43'
            FileSize: 525436
              Format: 'DICOM'
       FormatVersion: 3
```

Again, for space considerations, we do not produce the entire output and have truncated it after the first five fields of the **dcmInfo** structure. The data for any attribute within **dcmInfo** can then be accessed using dot notation:

```
% Using dot notation to access the "width" parameter
imageWidth = dcmInfo.Width
```

The default DICOM data dictionary supplied with MATLAB defines the DICOM attributes that MATLAB can recognize, including the associated DICOM tags and "VR" types. In most cases this default dictionary is sufficient for handling DICOM files. If an attribute is not recognized it is assigned as a *private attribute* and MATLAB attempts to interpret the data field. In cases where the private attributes cause problems, it is important to know that a new dictionary can be created and assigned. The **dicomdict** function will return the active DICOM dictionary when you use the 'get' option, and it also enables you to set the active DICOM directory using the 'set' option. Further details can be found in the MATLAB documentation for the **dicomdict** function.

4.2 IMPORTING IMAGE DATA

4.2.1 MATLAB's data model for DICOM images

This subsection concerns how MATLAB represents DICOM images. As per the DICOM standard, a DICOM file can be single-frame or multi-frame DICOM image. A single-frame DICOM image corresponds to a 2D image comprised of x- and y- coordinates–or columns and rows. A multi-frame DICOM image image adds an extra dimension. The number of frames can represent an additional spatial dimension (z-coordinate), or temporal dimension (time of the frame). A 3D data volume can also be represented as a set of single-frame files in a DICOM series, rather than as a multi-frame file. For example, a reconstructed CT volume is typically saved as a series of single-frame images.

MATLAB's DICOM functionality conforms to the DICOM standard, however, MATLAB adds an extra array dimension in all of the image types discussed above. This extra dimension sneaks in because of how MATLAB represents arrays of data that might contain color information. The color information in MATLAB is represented by *samples* or color channels. Different color samples, for example, red, green and blue (RGB)—occupy their own dimension. So, a grayscale DICOM image has one sample or color channel, whereas a true color DICOM image is comprised of three samples or color channels (RGB). With the samples or color channels as one of the dimensions, MATLAB's data model for 2D and 3D DICOM images is as depicted in Figure 4.1.

MATLAB data model for 2D DICOM images
A 2D DICOM image has x- and y-coordinates as columns and rows of the image. With the inclusion of color information, the image data has an additional dimension:

- MATLAB data model for 2D DICOM data $= rows \times columns \times samples$

MATLAB data model for 3D DICOM images
3D DICOM data in MATLAB is actually represented as 4D MATLAB array, where the third dimension is samples or number of color channels:

- MATLAB data mode for 3D DICOM data $= rows \times columns \times samples \times frames$ (here, *frames* can be slices along an additional spatial dimension or the temporal dimension)

There is a neat MATLAB function called **squeeze**, which removes any dimension of length 1. It returns an array with the same elements but with the singleton dimensions removed. This

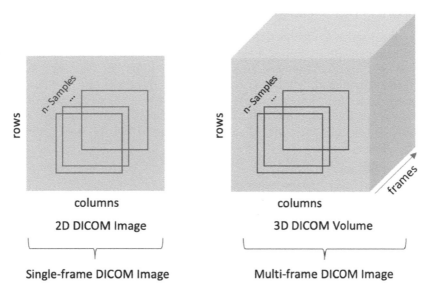

Figure 4.1: MATLAB data model for 2D and 3D DICOM images where *n-Samples* represents the number of color channels in the image.

is handy when importing grayscale DICOM data. For example, a 3D DICOM image stack of size $512 \times 512 \times 100$ will be imported into MATLAB as an array of size $512 \times 512 \times 1 \times 100$. Parsing this array into **squeeze** will return an array of size $512 \times 512 \times 100$.

4.2.2   Importing DICOM image data

Importing data refers to bringing the information from an external source into an application. In our workflow, importing data refers to loading the content of a DICOM file into the MATLAB workspace. DICOM data can be imported using **dicomread**. Example:

```
% Imports the DICOM data file US-PAL-8-10x-echo.dcm
dcmImage = dicomread('US-PAL-8-10x-echo.dcm');
```

The **dicomImage** variable could be a single frame grayscale image of size $m \times n \times 1$, a single frame true color image of size $m \times n \times 3$, or a multi-frame image of $m \times n \times 3 \times frames$.

The input DICOM file may be comprised of further attributes that aid in proper visualization of the DICOM image. One such example is the *colormap* which defines a mapping from a grayscale image representation to a color representation. The example below illustrates the syntax for extracting colormap from an input DICOM file:

```
[dcmImage,cmap] = dicomread('US-PAL-8-10x-echo.dcm');
```

The colormap associated with an image is an array of $k \times 3$ real numbers between 0 and 1. The three columns of the colormap matrix represent red, green and blue (RGB) values. The default number of rows (i.e., colors) in the colormap is 256. MATLAB maps the pixel value in an image to the corresponding red, green and blue values given by the colormap. Note that since colormaps represent RGB pixel values, the colormap will be empty for an image that is intended to be viewed as a grayscale image.

4.2.3 Importing DICOM image series

A DICOM image series is a collection of associated DICOM files with a common value for the `SeriesInstanceUID` attribute. A commonly encountered use for a series is to represent a 3D volume for CT or MRI images. In MATLAB, such 3D DICOM volume data could be imported using `dicomread`, a for-loop (to loop through slices) and the `sort` function (to order the slices correctly). However, it is much simpler to use the `dicomreadVolume` function. This function reads the individual DICOM files, performs the correct slice ordering of the individual data files, and constructs the corresponding MATLAB data array. Below is the syntax for its use:

```
% Read a DICOM volume
[dcmVolume,spatial,dim] = dicomreadVolume(source);
```

Above, `source` can be the name of the folder containing DICOM files, a string array of filenames or a cell array of character vectors containing file names. The variable `dcmVolume` is a MATLAB array consisting of the DICOM volume data; `spatial` is a structure specifying location, resolution and orientation of the slices in the volume; and `dim` is the dimension that has the largest offset from the previous slice (a scalar value: 1, 2 or 3, equating to the x, y or z dimension, respectively). To try `dicomreadVolume` using an MR data set supplied with MATLAB, first define `source = fullfile(dataDir,'dog')`.

4.3 WRITING AND ANONYMIZING DICOM DATA

4.3.1 Writing a DICOM file

We have seen how multidimensional DICOM data can be imported into MATLAB. Now we will see how to create a DICOM file. A new DICOM file can be created from scratch using a function called `dicomwrite`. Besides the image data, the `dicomwrite` function supports writing information such as the colormap and other metadata. Often it is convenient to create a new file based on the template of an existing file, to ensure that the correct metadata is present. The example below demonstrates how to read a DICOM image and write the image along with its metadata into a new DICOM file.

```
% Read knee1.dcm DICOM image
kneeImage = dicomread('knee1.dcm');

% Read metadata from Knee1.dcm
metadataOriginalImage = dicominfo('knee1.dcm');

% Write image kneeImage and metadataOriginalImage to a new DICOM file
dicomwrite(kneeImage, 'newKnee1.dcm', metadataOriginalImage);
```

The example copies the pixel data and metadata to the new file. These could have been modified prior to writing. While overwriting pixel values and many metadata fields poses no problem, those metadata fields consisting of *unique identifiers* (UIDs) require more care. New valid UIDs can be generated using the `dicomuid` function[2]:

```
% Create a new UID
newID = dicomuid;
```

This can be used to provide UIDs for DICOM attributes. New UIDs may be necessary, for example, when importing multiple copies of a study into the database of a software application. MATLAB fully implements three DICOM *Information Object Definitions* (IODs):

---

[2]Note: Every call to `dicomuid` always returns a new value.

Secondary Capture Image Storage (default), *CT Image Storage* and *MR Image Storage*. If the image being saved corresponds to one of these types, then the file written using `dicomwrite` should conform to the specification of the DICOM standard. If private attributes are present in the metadata they will not be written to the file unless the `WritePrivate` argument is set to `true`:

```
% Including private attributes when writing to a DICOM file
dicomwrite(imageData, 'newImage.dcm', metadata, 'WritePrivate', true);
```

If your image corresponds to an IOD type that has not been implemented in MATLAB (e.g., *Digital X-Ray Image*) then an attempt to write a file can be made by setting the `CreateMode` argument to 'Copy':

```
% Attempt to write to a DICOM file with an unsupported IOD
dicomwrite(imageData, 'newImage.dcm', metadata, 'CreateMode', 'Copy');
```

This should simply copy the data exactly, without verifying the metadata attributes or creating missing ones. However, in practice, errors may be encountered due to problems with type conversion of image or metadata for representation in MATLAB. If the in-built DICOM functionality of MATLAB proves insufficient, it can be extended by integrating with an external DICOM library. Please refer to Chapters 6 and 18 for examples of how to integrate with the *Evil DICOM* (.NET framework) and *etherJ* (Java) DICOM libraries, respectively.

4.3.2 Writing a DICOM series

It is possible to save a 4D MATLAB array (*rows* × *columns* × *samples* × *frames*) to a series of files through a single call to `dicomwrite` by specifying the `MultiframeSingleFile` argument as `false`. However, if a metadata structure is also supplied as an argument, identical metadata data will be written to every file in the series. This is generally undesirable, as some attributes will change from frame to frame, for example, `SliceLocation`. Therefore, typically, a DICOM series will be written by looping through all the frames of an image volume and applying `dicomwrite` repeatedly, specifying the appropriate metadata structure (from `dicominfo`) for each file. A patient can have multiple studies and each study can consist of multiple series. Each instance in a series shares a common value of `SeriesInstanceUID` and `StudyInstanceUID`. When writing a DICOM series it is therefore critical that these metadata fields are specified correctly. New UIDs can be generated using `dicomuid`:

```
% Copy the 22 files of the 'dog' series specifying new study/series UIDs
StudyInstanceUID = dicomuid;
SeriesInstanceUID = dicomuid;
for iFile = 0:21
    originalName = fullfile(dataDir,'dog',sprintf('dog%02d.dcm', iFile));
    dcmImage = dicomread(originalName);
    dcmInfo = dicominfo(originalName);
    dcmInfo.StudyInstanceUID = StudyInstanceUID;
    dcmInfo.SeriesInstanceUID = SeriesInstanceUID;
    newName = sprintf('copy%02d.dcm',iFile);
    dicomwrite(dcmImage,newName,dcmInfo);
end
```

4.3.3   Anonymizing DICOM data

Medical images typically contain identifiable information about subjects and some information which may not be directly related to a subject's identity, but may be able to help in recovering the identity. Hence, when using the set of medical images for training purposes, blind study or to show results of an analysis or study, it is required that this information be concealed or removed to help protect the privacy of the subject (see Chapter 9: *Regulatory considerations when deploying your work in a clinical environment*). However, researchers may want to retain specific fields or information in the data to be able to run statistical or other forms of detailed analysis. Therefore, the software workflow that needs to be used to anonymize or de-identify DICOM images will need to ensure it does not include any identifiable data in the exported files while retaining data that is needed.

Note that some DICOM attributes are mandatory (must be present), even if they are empty or filled with dummy information. Refer to the previous chapter (see Chapter 3: *Data sources in medical imaging*) for further information.

Anonymization and MATLAB

Anonymization in MATLAB conforms to DICOM Supplement 55 (*Attribute Level Confidentiality (including De-identification)*) and will modify or remove fields as needed to satisfy this standard[3]. It is the responsibility of the user, however, to ensure that anonymization is fully performed for their files (success may depend on file IOD and vendor implementation of the DICOM standard–in particular, check private attributes). It is also the user's responsibility to check that the de-identification satisfies any further requirements on Protected Health Information in their jurisdiction.

The function **dicomanon** is used for removing confidential medical information from a DICOM file while retaining the image data and other attributes. To create a version of a DICOM file with all the personal information removed, the following syntax can be used:

```
% Create a version of a DICOM file with all personal information removed
dicomanon('US-PAL-8-10x-echo.dcm','US-PAL-anonymized.dcm');
```

However, to create a version of a DICOM file with the patient's personal information removed, while retaining certain fields that could be useful like the age, gender, etc., we can specify the fields in the syntax as shown below:

```
% Remove only patient's personal information while retaining certain fields
% such as age, gender, etc.
dicomanon('US-PAL-8-10x-echo.dcm','US-PAL-anonymized.dcm','keep',...
        {'PatientAge','PatientSex','StudyDescription'});
```

Note that the anonymization generates new UIDs for some attributes. This means that a set of anonymized images in a series will no longer have common study and series UIDs unless you explicitly define them. This can be done in the following manner:

```
% Anonymize the 22 files of the 'dog' series specifying new study/series UIDs
values.StudyInstanceUID = dicomuid;
values.SeriesInstanceUID = dicomuid;
for iFile = 0:21
    originalName = fullfile(dataDir,'dog',sprintf('dog%02d.dcm', iFile));
    newName = sprintf('anon%02d.dcm',iFile);
    dicomanon(originalName,newName,'update',values);
end
```

---

[3]Supplement 55 can be found at: https://www.dicomstandard.org/News/ftsup/docs/sups/sup55.pdf

4.4 VISUALIZATION

So far in this chapter we have learned to import single-frame and multi-frame DICOM data into MATLAB and how to modify it and save it. Often, however, we also wish to visualize images. Importing and displaying DICOM data opens another door to further analyze the images. This section covers different techniques to visualize 2D and 3D volumes. First the typical steps for visualizing a DICOM data will be discussed followed by the techniques to visualize DICOM data. There are many available functions to perform visualizations and an exhaustive list will not be presented here: the reader is encouraged to explore the MATLAB documentation.

Step 1: Knowing the data: Before visualizing it is very important to understand the data. Functions such as `dicomCollection` and `dicomdisp` can be useful for examining properties such as the modality, dimensionality and size of the data. Knowing the values and ranges is also essential in choosing the correct manner in which to display data and this information can be extracted using `dicomread` and `dicomreadVolume`.

Step 2: Select appropriate plotting routine: MATLAB provides a range of plotting routines for visualizing image data. The appropriate choice will depend on the dimensionality of the data, the data type and the features that need to be investigated. MATLAB provides functions for displaying single 2D images (e.g., `imshow` and `imagesc`), stacks of 2D images (e.g., `montage`, `sliceViewer` and `orthosliceViewer`) and 3D volume display (`volshow`).

Step 3: Defining the representation: A mapping is required to represent the pixel values on the screen. For a 2D scalar image, this can be a simple as defining the window width and level, or a colormap. For the visualization of data in 3D the factors to consider may be more numerous, including orientation, coloring, alphamap (transparency) and lighting.

4.4.1 Displaying 2D data

A 2D or single-frame image can be displayed using the `imshow` or `montage` functions. These functions are part of the MATLAB's *Image Processing Toolbox*. The key difference between `montage` and `imshow` is that `montage` allows the display of a stack of 2D images (multiple frames), whereas `imshow` can only display a single frame.

Display an image stack using `montage`
The function `montage` displays multiple image frames in an array as a rectangular montage. By default the display is arranged to form a square. The code snippet below displays the echocardiogram ultrasound image *US-PAL-8-10x-echo.dcm*:

```
% Load 'US-PAL-8-10x-echo.dcm' DICOM image and display it using 'montage'
[dcmImage,cmap] = dicomread('US-PAL-8-10x-echo.dcm');
figure;
montage(dcmImage)
```

By default, `montage` applies a grayscale map to the displayed image. This is shown in Figure 4.2. The colormap can be modified after the call to `montage` by passing a custom colormap to the `colormap` function:

```
colormap(cmap)
```

Alternatively, the colormap can be passed to `montage` as an argument. It is also possible

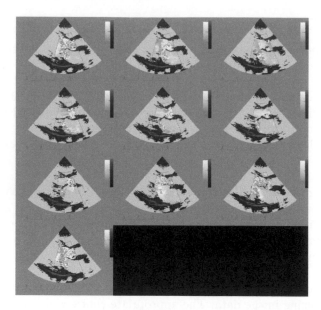

Figure 4.2: `dicomImage` visualized using the `montage` function.

to pass other arguments to `montage`. The following code snippet passes the colormap `cmap` returned from the dicomread function to `montage` and displays the returned image as a [2,5] rectangular grid. Figure 4.3 shows the output figure.

```
% Use colormap 'map' returned in the above step to display the DICOM imaged
% as [2,5] grid in a rectangle
figure;
montage(dcmImage, cmap, 'Size', [2,5])
```

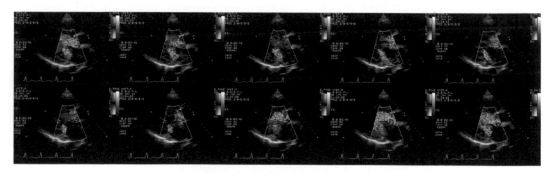

Figure 4.3: `dicomImage` visualized using `montage` with custom colormap in[2, 5] rectangular grid.

By default montage displays all the images from an image stack. Specific images from a stack can be displayed by passing *indices* as an argument to montage. For example, the code snippet below will display frames 3:5 from the image stack:

```
% Specify indices to display certain frames
figure;
montage(dcmImage, cmap, 'Indices', 3:5)
```

Since MATLAB version R2019b, two new functions have been available for viewing stacks of 2D images: `sliceViewer` and `orthosliceViewer`. The former function displays a single frame of a multi-frame image, but has a slider to dynamically select the slice viewed. The latter function displays three images of a multi-frame image corresponding to the three orthogonal planes.

Display a single frame image using `imshow`

A 2D image or an individual frame of a DICOM image stack can also be displayed using `imshow`. The imported image in the example of the previous subsection has 10 frames. The code snippet below extracts the first frame from the image stack and displays the image using `imshow`.

```
% Get the first frame and display firstFrame using imshow
firstFrame = dcmImage(:,:,:,1);
figure;
imshow(firstFrame, cmap);
```

The resulting graphic is shown in Figure 4.4.

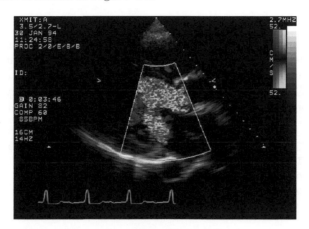

Figure 4.4: The first frame of `dicomImage` visualized using `imshow` function.

Similarly, other frames can be extracted by specifying a different index to the `dicomImage` array. As with `montage`, `imshow` uses a grayscale colormap by default. In the example above, we passed `cmap` as an input parameter. Unlike with `montage`, however, a single frame grayscale DICOM image can be displayed directly using `imshow`, without using `dicomread`.

```
% Display single frame image directly using imshow
figure;
imshow('CT-MONO2-16-ankle.dcm','DisplayRange',[])
```

Here, setting `DisplayRange` to `[]` scales the display of the 16-bit values to span the full display range.

The intensities in DICOM images are typically saved as unsigned integers and might be rescaled and/or shifted to fit the encoding of the pixel data tag content. To obtain the real or original values you should apply the *Rescale Intercept* (0028,1052) and *Rescale Slope* (0028,1053) tags. This is necessary for CT images to convert the pixel intensities to Hounsfield units. However, it is a good practice always to apply these tags, if they are present. Here is an example of how this is done in MATLAB using a CT image of a test object[4].

---

[4]Available as the file "phantomImage.dcm" in the book Git repository: `https://bitbucket.org/DRPWM/code`

```
% For CT: convert pixel values to HU-values
CTInfo = dicominfo('phantomImage.dcm'); % Import metadata
CTImage = double(dicomread('phantomImage.dcm')); % Import image & convert to double
CTImageRescale = CTInfo.RescaleSlope.* CTImage + CTInfo.RescaleIntercept;
```

The conversion to a floating point data type (e.g., *double*) is necessary for the rescaling operation. The rescaled and offset image can be displayed as follows:

```
% Display image data with standard window leveling (0 = black, 1 = white)
figure;
imshow(CTImageRescale);
```

Displaying such imported DICOM images in MATLAB, you will notice that if you use the standard windowing and levelling, you will not see much of the image. This is because the standard window levelling applies black to value 0 and below, and white to value 1 and above (see Figure 4.5a). You can instead apply the full intensity span as the window width, as below:

```
% Display image data with "full" window leveling (min to max)
figure;
imshow(CTImageRescale,[]);
```

Unfortunately, an image visualized in this manner will tend to lack contrast. This is because this method rescales the image to have the lowest value in the image as black and the highest as white, and most values in a typical DICOM image lies somewhere in between (see Figure 4.5b). Generally, when looking at DICOM images you want to start by applying the value suggested in the *Window Center* (0028,1050) and *Window Width* (0028,1051) tags, if present:

```
% Display image data with window leveling defined in the DICOM metadata
windowCenter = CTInfo.WindowCenter(1);
windowWidth = CTInfo.WindowWidth(1);
lowerBound = windowCenter - windowWidth/2;
upperBound = windowCenter + windowWidth/2;
figure;
imshow(CTImageRescale,'DisplayRange',[lowerBound upperBound]);
```

This provides a much more useful visualization, as demonstrated in Figure 4.5c.

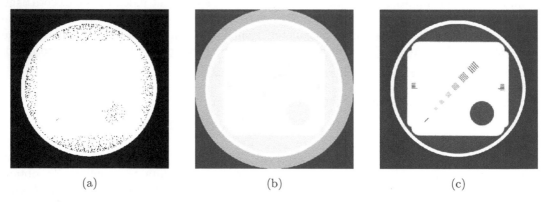

(a) (b) (c)

Figure 4.5: A DICOM CT image of a phantom displayed using three different window settings: (a) MATLAB default window leveling, (b) full range of the image intensities, and (c) the *Window Center* and *Window Width* tags.

4.4.2 Displaying 3D volume data

The appropriate technique for visualizing volume data depends on the type of the data at hand. A volume data set can either be a multidimensional vector volume data or scalar volume data. A vector volume data contains two or more values per voxel, defining the components of the vector. A scalar volume data set contains a single value (or color) per voxel. A typical DICOM volume data set is a scalar volume, which can be visualized in MATLAB using various plotting functions, such as: `volshow`, `isosurface`, `slice` and `contourslice`.

Displaying a volume using the *Volume Viewer* app

3D DICOM volume data can be visualized using the *Volume Viewer* app instead of using an equivalent command line function (`volshow`). 3D volumetric data can be visualized interactively using the *Volume Viewer* app and this will be explored further here. The following example illustrates loading, visualization and exploration of a DICOM volume data using the app. We will again use MATLAB's 3D MR data set called *dog*.

Step 1: Load the 3D DICOM volume data

Let us begin with loading the DICOM dog data into MATLAB's workspace.

```
% Step 1- Load DICOM volume data and extract information from the data
[dicomVolume,spatial,dim] = dicomreadVolume(fullfile(dataDir,'dog'));
```

`dicomVolume` is the returned 4D MATLAB array containing the DICOM volume. Note that this is actually 3D data, however, as discussed in Section 4.2.1, it additionally contains a *samples* dimension. The size of this data is $512 \times 512 \times 1 \times 22$, that is, there are 22 image slices each with the size of 512 rows and 512 columns. This data set has a single sample value per pixel and therefore represents a grayscale image. We can find additional information about this data set by using the `dicomCollection` or `dicomdisp` functions. We will use the `squeeze` function to remove the singleton dimension:

```
% Remove singleton dimension from the data set
dicomVolume = squeeze(dicomVolume);
```

Step 2: Open the Volume Viewer app

The Volume Viewer app can be opened in two ways:

- MATLAB command prompt: enter `volumeViewer`
- MATLAB's toolstrip: Open *Apps* tab, click the *Volume Viewer* app icon under *Image Processing and Computer Vision* section

Step 3: Load data into the app

In the *Volume Viewer* app load data by clicking on *Load Volume*. The volume data can be imported either from file, from a DICOM folder containing DICOM images, or also from workspace. In this example we will import the `dicomVolume` variable from the workspace using the *Import from workspace* option under *Load Volume*. The *Variables* table in the *Import From Workspace* window will show all of the valid 3D volume data of the size M-by-N-by-P if there are multiple 3D volume data present in the current MATLAB workspace.

Step 4: Explore the data

The volume viewer app provides several options that can be explored in the app. These options are:

- **Background color**: The background color of the display window can be changed by clicking on *Background Color* in the *Layout and Background* option of the app. For this example, we will go with the default background color.

- **Zoom and Rotate**: It is possible to zoom and rotate the image to explore the volume. The spatial position of the x, y and z axes can be seen as you rotate the image in the *Orientation Axes* window of the app.

- **Change View**: By default the data is displayed as a *Volume* in the app. This view can be changed to *Slice Planes* by clicking on the *Slice Planes* in the *View* panel. Once the slice planes view is enabled, individual slices can be viewed by scrolling on the *XY Slice*, *XZ Slice* and *YZ Slice* windows of the app.

- **Changing Spatial References**: By default the volume is displayed such that it appears compressed along the z-axis. This volume data can be better visualized by changing the spatial scale, that is, by specifying the number of units per voxel in X, Y and Z dimensions. For example, set 8 units/vx for the *Z-axis* to change the view to a cube.

- **Refine View with Rendering Editor**: The view of the volumetric data can be further refined by using the rendering editor. The rendering editor provides options for:

  - Selecting the methods for volume viewing. Options include: *volume rendering*, *maximum intensity projection* and *isosurface*. For our example we will go with *volume rendering* which is also the default volume viewing method in the app.

  - Specifying alphamap. Specifying an appropriate alphamap is crucial part of the volume visualization. By choosing an appropriate alphamap, the structure of interest can made opaque keeping other structures transparent. The default alphamap is *linear*. Additionally, the app provides some presets for alphamap based on the type of data. If the input data is a CT or MRI, one can select a corresponding CT or MRI presets for alphamap. For example, if the input data is a CT scan of a lung, selecting *ct-lung* for the alphamap gives better visualization. The alphamap can also be customized by modiyfing the curve of the plot. For the data in our example, we will select the alphamap to be *mri-mip*.

  - Specifying colormap. Besides the generic MATLAB colormaps, the app also provides colormap presets based on the type of image. One can also apply a custom colormap. For our example we will select colormap value to *mri*.

  - Changing lighting. Lighting can be turned on/off from this setting.

A screenshot of the resulting view is presented in Figure 4.6. These are not the only the settings available to visualize this data. We suggest the readers to explore further by trying out other presets for alphamap, colormap, different viewing angles and other settings in the *Volume Viewer* app.

Step 5: Export and reuse the settings
The rendering and camera configuration settings can be exported to a structure by clicking on the *Export* button in the app. By default the name of the structure is "config". This configuration file can be used to recreate the view we created in the app, using the function `volshow`. Use the syntax below:

```
% V is a 3D grayscale volume
% config is a struct containing exported rendering information
volshow(V,config);
```

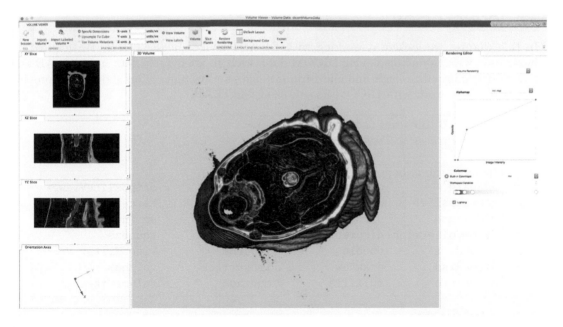

Figure 4.6: The DICOM MR data set *dog* visualized in the *VolumeViewer app*.

4.4.3 Displaying data with a time dimension

A 3D data set consisting of 2D spatial and a time dimension can be visualized using `montage`, with each image in the grid corresponding to a time point. Alternatively, the time dimension can be scrolled through when using the `sliceViewer` function. A 4D data set, consisting of a 3D volume and a time dimension, requires a new approach. One approach to such data sets (or any with a time dimension) is to use a *for-loop* to iterate through the frames or volumes, calling a plotting function each time. In such cases, the function `drawnow` is useful if you want to see the updates on the screen immediately as the for loop progresses. A 3D example (2D slice plus time) of this is demonstrated in Chapter 20 (*Importation and visualization of ultrasound data*), where the `imagesc` plotting function is used. Similarly, 4D data (3D volume plus time) can be visualized using the `volshow` function in a loop.

An alternative approach is to create a video file, which can be done using the `VideoWriter` or `imwrite` functions. For 3D (2D plus time), this is simple, as each 2D frame will correspond to a frame of the the video. For 4D data (3D volume plus time), 2D views of the 3D volume must be generated. One way to do this is generate a graphic using `volshow` and extract the view using `getframe`. The reader should refer to the MATLAB documention of `volshow` for more information.

4.5 CONCLUSION

In this chapter we looked at working with DICOM images in MATLAB. We focused on importing, writing and displaying DICOM data, ways to anonymize data in MATLAB and techniques to visualize single and multi-frame images and volumes. This should help provide a good basis to work with DICOM images using MATLAB.

MATLAB toolboxes used in this chapter:
Image Processing Toolbox

Index of the in-built MATLAB functions used:

addpath	dicomread	getframe	orthosliceViewer
contourslice	dicomreadVolume	imagesc	sliceViewer
dicomanon	dicomuid	imshow	sprintf
dicomCollection	dicomwrite	imwrite	squeeze
dicomdict	double	isdicom	VideoWriter
dicomdisp	drawnow	isosurface	volshow
dicominfo	fullfile	montage	volumeViewer

Acknowledgements

Piyush Khopkar wishes to thank his family for their constant encouragement. He is indebted to number of individuals at MathWorks for being supportive of this book project, proofreading chapter draft and providing valuable feedback. In particular, Annapoorna Mahalingam, Chris Portal, Jeff Mather, Josephine Dula, Linh Le, Monica Wang and Nakul Khadilkar. Lastly, the editors for giving the opportunity to contribute in this book.

Viju Ravichandran wishes to thank MathWorks for being supportive of this book project and various engineers at MathWorks for providing constructive feedback. Viju would also like to thank Johan Helmenkamp for his continuous support through the process and being flexible with the timeline of submission of drafts.

## CHAPTER 5

Creating automated workflows using MATLAB

Johan Helmenkamp

Medical Radiation Physics and Nuclear Medicine, Karolinska University Hospital, Stockholm, Sweden

Sven Månsson

Medical Radiation Physics, Department of Translational Medicine, Lund University, Skåne University Hospital, Malmö, Sweden

CONTENTS

O NE of the strengths of wielding programming skills is to be able to automate repetitive tasks. This chapter will demonstrate how such automated processes can be set up using a simple example based on the calculation and reporting of the signal-to-noise ratio (SNR) for a flat-field x-ray image.

5.1 INTRODUCTION

Tasks that involve some form of computational activity and are repetitive are likely to be highly suitable for automation. As computers are massively more efficient than humans for tasks that can be reduced to a computation, automation offers significant benefits. Automation cancels the need to wait for human input and eliminates the risk of errors and inconsistencies caused by human operators. Human resources can be freed up to focus on other activities where computers are yet to prevail. Even relatively simple tasks, such as calculating the SNR in some region-of-interest (ROI) of an image, may benefit in terms of efficiency and consistency from automation. A simple overview of the process components of an automated workflow is showed in Figure 5.1. The exemplified components may of course vary according to the specifics of each coding project.

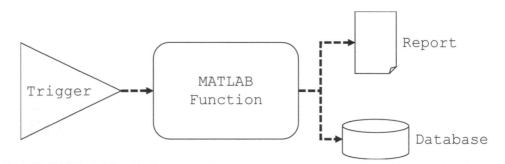

Figure 5.1: Schematic overview of an automated process in which some core MATLAB function or program is triggered by an event, and where the output is automatically reported and stored in a database.

A few examples of triggers are listed below, together with typical ways in which data can be accessed and imported and what kind of outputs are commonly produced.

Triggers

Besides the human operator as a trigger of automatic processes (like turning on the ignition in a car), the triggering event may often be initiated by another computer program, or hardware such a sensor. In the field of Radiology, a common trigger is a DICOM server that is scripted to initiate your program when certain conditions are met, such as the arrival of specific phantom or patient images. Another typical trigger can be a clock that initiates a program at regular time intervals. There are several conceivable events that could work as triggers in any given context.

Importing data

Once your program is triggered, it is likely going to need some data to crunch. The input data can be in a wide variety of forms. Of course, in Radiology the input data is not seldom a stack of DICOM images. The triggering event (like a DICOM server), can provide the path to the DICOM files as part of the triggering event or a standardized path can be used, although the latter is not as elegant. Other typical data sources are *databases* of various sorts (one common type is introduced in this chapter, namely SQL, or *Structured Query Language*). Databases are essentially just a type of container for data. Once triggered, your program can poll one or several databases for some data such as patient information or base measurement values, and push new data back to them. It is not uncommon that healthcare information systems, like a RIS (Radiology Information System), store their data in SQL-type databases.

Typical outputs

The output of an automated process of course varies significantly depending on the particular application. In image processing applications you will expect some form of quantified entity to emerge, such as measurements based on the input data. These could be anything from a single number to a gigantic set of extracted radiomic features or AI-based classifications of the information content. It is worth mentioning that an automated process does not need to have any quantifiable output; it may simply be a process that is designed to re-arrange a folder structure, move files around or trigger other events.

Showcasing example features of an automated process
A simple case study is used in this chapter to exemplify the concepts of automation. The sections below will detail how to extract image data from a flat-field image acquired at a projection radiography clinic, how to calculate the SNR in a centrally placed ROI and examples of how to report and store the results. The image was acquired using the IEC RQA-5 standard x-ray energy spectrum[15] at a detector air kerma level of approximately 3 μGy.

It is worth noting that the MATLAB code presented in this chapter can readily be expanded for use in other applications, such as linearization of x-ray images (conversion of pixel values to pixel detector air kerma levels), or as part of an image-based linearity test of an x-ray generator (linearity of detector output versus generator tube current-time product)[16]. The concept of extracting information from a ROI in x-ray images is also central to the calculation of the Noise Power Spectrum (NPS), and many other metrics of interest[17]. In other words, it is quite useful to master the basics of the code in this chapter.

5.2   MANUAL CALCULATION OF SNR

The manual process of calculating SNR involves a few different steps (see Figure 5.2). Whatever software is used for manually extracting metrics from the image (e.g., the freely-available ImageJ program), it has to be configured to output the mean value and standard deviation for its measurement.

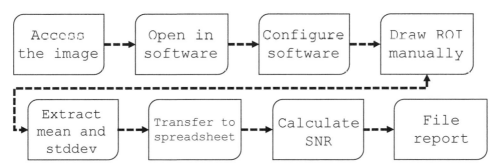

Figure 5.2: Overview of an example of a manual process for calculation and reporting of the SNR in one image. There may be more steps involved, especially for the first activity, but for simplicity this suffices for illustration of the major components.

After manually defining the ROI (targeting a central placement and about 100×100 pixels in area), the values are transferred to a spreadsheet program, where calculation of the SNR according to Equation 5.1 can be performed.

$$\text{SNR} = \frac{\bar{s}}{\sigma} \tag{5.1}$$

where $\bar{s}$ and σ are the mean and standard deviation of the pixel values in the selected ROI of the image data, respectively. An average human operator should be able to go through the above described process in about 1–2 minutes, depending on the software used. 1–2 minutes is a long time for extracting such a simple metric from a single image. The duration is further lengthened by the number of images, the number of ROIs per image, of which there may be many depending on the measurement, and the number of times of the process needs repeating. Such a task could quickly become laborious for the average human operator and perhaps even overwhelming and impractical for more sophisticated measurements.

5.3 AUTOMATING THE SNR CALCULATION USING MATLAB

When designing a program that is going to be a part of an automated workflow there are more things to consider than just writing the code for the calculation of the metric in question. The code has to be written so that it will work as a component in the intended process. In this example a program called `calcSNR` is designed to look for a single DICOM file in the current working directory, or terminate otherwise. The entire workflow is visualized in Figure 5.3.

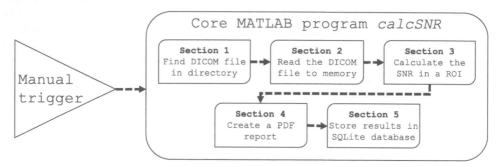

Figure 5.3: The five main sections of the core MATLAB program `calcSNR` executes automatically and sequentially in order once triggered. For simplicity, the trigger will be manual in this case.

How the program should be packaged is another aspect to consider. It can be a MATLAB function or app that has to be run inside MATLAB, or it can be deployed as a stand-alone executable (*.EXE) using the *MATLAB Compiler*. Compilation to an executable has the advantage of making it possible to automatically trigger and run the process on a computer that does not have an installation of, or license for, MATLAB. Another possibility that probably offers the most light-weight and fastest executable programs is to use *MATLAB Coder*, that generates C or C++ code directly from MATLAB code. The C/C++ code can thereafter be packaged into an executable. For more information about how to create MATLAB apps and deployment strategies, please refer to Chapter 8 (*Sharing code*). In more advanced environments it may be beneficial to *containerize* your program for deployment. Please see Chapter 18 (*Image processing at scale by "containerizing" MATLAB*) for a case-study in this.

5.3.1 Code section 1: Finding the DICOM image file

The first step is to find the path to all files in the current working directory. For this example, a neat little function called `dirPlus` developed by Kenneth Eaton is used for this purpose and it returns the paths for all files in the directory as a cell array[1]:

```
function calcSNR(~)
%% Section 1 - finding the path to the DICOM file

% Use the function dirPlus to create a list of all files in the current
% directory and subfolders
% "pwd" is a MATLAB command that returns the current working directory
allPaths = dirPlus(pwd);
```

Once the paths to all files are known, one way to single out which one is the DICOM file is to use the function `isdicom`:

---

[1]The function `dirPlus` is available for download on MATLAB File Exchange.

```
% Pre-allocate the vector indexDICOM in computer memory
indexDICOM = zeros(numel(allPaths),1);

% Loop through the paths of all files in the current directory and
% sub-folders and create an index of all DICOM file locations in allPaths
for i = 1:numel(allPaths)
    imagePath = char(allPaths(i));
    if isdicom(imagePath)
        indexDICOM(i) = 1;
    end
end

% Close the program if there are no DICOM-files in the current directory,
% or if there are more than one
if sum(indexDICOM) == 0 || sum(indexDICOM) > 1
    return
else
    % Remove all paths from allPaths which are not DICOM-files
    pathImage = allPaths(logical(indexDICOM));
end
```

It is also a good idea to define the paths to locations where the reports and database will be located:

```
% Define paths to folders containing the reports and database
reportPath = fullfile(pwd,'Reports');
databasePath = fullfile(pwd,'Database');

% Create the report and database paths if they don't exist
if ~exist(databasePath, 'dir')
    mkdir(databasePath);
end
if ~exist(reportPath, 'dir')
    mkdir(reportPath);
end
```

5.3.2   Code section 2: Reading the DICOM file

Now that the path to the DICOM file is known, it is straight forward to read the content of the file into computer memory using the functions dicomread and dicominfo:

```
%% Section 2 - reading the contents of the DICOM file to memory

% Read the metadata and the image data. Notice the conversion to datatype
% "single" for the image data!
dcmInfo = dicominfo(pathImage{1});
dcmImage = single(dicomread(pathImage{1}));
```

We continue by defining the ROI size, and calculating its intended position in the image based on the image size in pixels across the x- and y- dimensions:

```
% Find the size of the image in pixels and define the size of the ROI
sizeImage = size(dcmImage);
sizeY = sizeImage(1);
sizeX = sizeImage(2);
roiSize = 100;

% Define the coordinates of the ROI, so that it'll be placed in the center
% of the image, based on the image size
ROIPositionxMin = round(sizeX/2 - roiSize/2);
ROIPositionxMax = round(sizeX/2 + roiSize/2 - 1);
ROIPositionyMin = round(sizeY/2 - roiSize/2);
ROIPositionyMax = round(sizeY/2 + roiSize/2 - 1);
```

5.3.3   Code section 3: Calculating the SNR

With the image data in your computer's memory and an ROI defined, the mean value and standard deviation of the image data within the ROI can be calculated, which brings us to Section 3 and the calculation of the SNR metric according to Equation 5.1:

```
%% Section 3 - calculate the SNR in the ROI

% Calculate the mean and standard deviation in the ROI
meanVal = mean2(dcmImage(ROIPositionyMin:ROIPositionyMax,...
    ROIPositionxMin:ROIPositionxMax));
stdVal = std2(dcmImage(ROIPositionyMin:ROIPositionyMax,ROIPositionxMin: ...
    ROIPositionxMax));

% Calculate the SNR, which is casted to double for compliance with the
% SQLite database later on
SNR = double(meanVal/stdVal);
```

Note that the definitions of the ROI shape, size and position are written explicitly in the code (know as hardcoding). For examples of more versatile management of ROIs using MATLAB, please see Chapter 11 (*Automating Quality Control tests and evaluating ATCM in Computed Tomography*).

Using the above MATLAB code in combination with the execution timing functions tic and toc it is possible to time how long it takes to run the code. Sections 1 through 3 runs from start to finish in about 200 milliseconds on a standard laptop—over a hundred times faster that the manual workflow described in Section 5.3.2.

5.3.4   Code section 4: Reporting the results

MATLAB offers many options to document and report data. MATLAB can for instance connect to a Microsoft Word document using the function actxserver, populate it with data in tables, inserting figures or graphs and then have Word save the file in any supported document format (such as PDF). The report can be automatically emailed to one or more recipients using the MATLAB function sendmail. For this example, the MATLAB Report Generator will be used[2]. The MATLAB Report Generator includes functionality for direct generation of reports in PDF, Word, PowerPoint and HTML, based on an internally generated and editable report template. See Figure 5.4 for an example PDF report that has been generated by the code in this section.

---

[2]The reader is advised to consult the MATLAB documentation for more information.

Signal-to-noise ratio report

SNR: 117.6

Exposure parameters
kVp: 70 kV
mAs: 8 mAs

Figure 5.4: A section of a PDF example report generated by Section 4 of `calcSNR`, utilizing the functionality of the MATLAB Report Generator. The `root.css` style sheet has been edited to use only black-and-white for all attributes, and with no line borders for tables.

The report template `myTemplate` can be created with the following commands:

```
% Import the Document Object Model (DOM) base class
import mlreportgen.dom.*

% Create a copy of the default DOM template file
mlreportgen.dom.Document.createTemplate('myTemplate','pdf');

% Unzip the template file so that it can be edited manually
unzipTemplate('myTemplate.pdftx','myTemplateDirectory');
```

Modification of the template is usually made by editing the file `stylesheets/root.css` in the unzipped directory, using a text editor. Once edited, the modified template is created by the command:

```
zipTemplate('myTemplate.pdftx','myTemplateDirectory');
```

The above only needs to be done once. For deployment, the Report Generator is initialized in the following way, after which a report object can be created:

```
%% Section 4 - report the results

% Fire up the MATLAB Report Generator and make it available for deployment
import mlreportgen.dom.*
makeDOMCompilable();

% Define the filename of the report based on DICOM metadata
reportFileName = erase([dcmInfo.Manufacturer '_' dcmInfo.ManufacturerModelName ...
    '_' dcmInfo.StudyDate],'""');

% Create the Document object and create some headings
report = Document(fullfile(reportPath,reportFileName), 'pdf','myTemplate');
open(report);
append(report, 'Signal-to-noise ratio report','Title');
```

It would be nice to visualize the image with the predefined ROI in the report. In order to append an image to a report object, it first has to be created and saved to some directory path. In this case, the report path and the PNG file format at the current screen resolution will do just fine:

```
% Display the image in the background, visualize the ROI and save the image
% into the working directory
figSNR = figure('Visible','off');clf;
set(figSNR, 'position', [65 99 512 512])
ax = axes(figSNR, 'position', [0 0 1 1]);
imagesc(dcmImage);
colormap(gray);
axis off
rectangle('Position',[ROIPositionxMin ROIPositionyMin roiSize roiSize], ...
    'LineWidth',2);
print(figSNR,'-r0',fullfile(reportPath,'figSNR.png'),'-dpng')
```

After the image is created and saved, it can be appended to the open report object. In this case, it will be placed inside the first row and first column in a table object, together with the result of the SNR calculation in the second column:

```
% Insert the image and calculation result in the Document object
img = Image(fullfile(reportPath,'figSNR.png'));
img.Width = '5cm'; img.Height = '5cm';
result = Paragraph(sprintf('SNR: %1.1f', SNR),'Bold');
table = Table({img,result}); table.Width = '100%';
table.TableEntriesVAlign = 'middle';
table.entry(1,1).Style = {Width('6cm')};
append(report, table);
```

It could also be helpful for documentation purposes to include some information about the relevant exposure parameters. The example code below creates a table and populates it with the kV and mAs values from the DICOM metadata.

```
% Some exposure parameters
p1 = Paragraph('kVp: ');p1.Style = {WhiteSpace('preserve')};
d1 = Paragraph([num2str(dcmInfo.KVP) ' kV']);
d1.Style = {WhiteSpace('preserve')};
p2 = Paragraph('mAs: ');p2.Style = {WhiteSpace('preserve')};
d2 = Paragraph([num2str(round(dcmInfo.ExposureInuAs/1000),1) ' mAs']);
d2.Style = {WhiteSpace('preserve')};

% Present the parameters in the report
append(report, Heading3('Exposure parameters'));
exposureTable = Table({{p1;p2},{d1;d2}});
exposureTable.entry(1,2).Style = {Width('6cm')};
append(report, exposureTable);
```

Once populated with all necessary information, the report object is closed and the PNG copy of the image is deleted:

```
close(report);

% Delete the separate image file
delete(fullfile(reportPath,'figSNR.png'));
```

The report can be viewed in a PDF document viewer such as Adobe Acrobat Reader, or from within MATLAB using the `rptview(report)` command.

5.3.5   Code section 5: Storing the results in a structured database

Working with databases does not have to be complicated. In fact, even simple and small databases offers several advantages over spreadsheets. Some general advantages of using databases are:

- Access to the data can be controlled through the use of accounts
- The data integrity can be protected from tampering by limiting the rights associated with the accounts (such as *read-only* rights)
- Databases use standardized syntax for commands, such as doing data queries or data inputs
- Databases can store huge amounts of data while still being highly manageable
- Databases offers the potential to work more efficiently and centrally with data management, and the data can more easily be shared with peers
- Some databases automatically do backups of the data

MATLAB offers several ways to both setup and connect to external databases of various kinds. For this example, a simple SQLite[18] database will be used to demonstrate the basic workflow and syntax. In the MATLAB Database Toolbox, SQLite does not require you to setup an external database or extra drivers, making it simple to get started quickly.

We start by defining the filename of the database file:

```
%% Section 5 - store the results in an SQlite database

% Define the database filename
dbFile = fullfile(databasePath,'DB.db');
```

An SQLite database has a simple, structured format. It is like a table, with columns and rows. For this example, we will consider a database with five columns; StudyDate, SNR, InstitutionName, StationName and SOPInstanceUID. We will define the names of the columns and populate a cell array with these, and populate another cell array with the data intended for storage in the database[3]:

```
% Define the column names and data to input into the database table
colNames = {'StudyDate','SNR','InstitutionName','StationName','SOPInstanceUID'};
colData = {dcmInfo.StudyDate,SNR,dcmInfo.InstitutionName, ...
        dcmInfo.StationName,dcmInfo.SOPInstanceUID};
```

The following **if** statement creates a new database file, if it does not already exist, and inputs the data. If the database file already exists, the code under the **elseif** statement inputs the data only if it has not already been previously stored. The latter is to avoid having multiple entries in the table for the same image:

```
% Get a list of all file paths in the databasePath
fileIndex = dirPlus(databasePath);
sizeIndex = size(fileIndex);
```

(code continues on next page)

---

[3]SQL databases allow you to also store files, like a PDF report, in the database as a so called BLOB data type (*Binary Large OBject*). However, this is currently not supported in the SQLite implementation in MATLAB.

```
% Run the following code only if there are no files in the database path
% - this creates a new database file and populates it with the data
if sizeIndex(1) == 0
    conn = sqlite(dbFile,'create');
    createTable = ['create table SNRTable ' ...
    '(studyDate VARCHAR, SNR DOUBLE, InstitutionName VARCHAR, ' ...
    'StationName VARCHAR, SOPInstanceUID VARCHAR)'];
    exec(conn,createTable)
    insert(conn,'SNRTable',colNames,colData)

% Otherwise, run the following code only if there is exactly 1 file in the
% database path and the filename exactly matches the expected
% - this connects to the database file and populates it with the data
elseif sizeIndex(1) == 1 && strcmp(fileIndex{1,1},dbFile)
    conn = sqlite(dbFile);

    % Check if the data already exists in the database, and write to the
    % database only if not
    sqlQuery = 'SELECT `SOPInstanceUID` FROM SNRTable';
    results = fetch(conn,sqlquery);
    if strcmp(results{1,1},dcmInfo.SOPInstanceUID)
        close(conn)
        return
    else
        insert(conn,'SNRTable',colNames,colData)
    end
end
```

Finally, we close the connection to the database:

```
% Close the connection to the database
close(conn);
end % Ends the function
```

The code for the program **calcSNR** described above totals around 2-3 seconds from start to finish on a standard laptop. This is considerably faster compared to the 1-2 minutes it will take for the average human operator to perform these calculations manually.

5.3.6   Final touches: Deploying the code

As previously mentioned, MATLAB offers several different options for application deployment. To create a MATLAB app installation file you can select *Package App* under the *APPS* tab in MATLAB main window. This produces an .mlappinstall file that can be distributed to another computer with MATLAB installed. The most straight forward option to create a *stand-alone* application, however, is to opt for the *MATLAB Compiler*. It is easy to use and guides you through the different steps. This is what we will be using for the "calcSNR" program that we built in this chapter. Again, please refer to Chapter 8 (*Sharing Code*) for further information on application deployment.

The MATLAB Compiler can be launched by navigating to the *APPS* tab and clicking on *Application Compiler* in the Apps Gallery list. The compiler allows you to define what is to be packaged together with other options. Start by adding the "calcSNR.m" as the *main file*. MATLAB Compiler automatically detects any dependencies to other functions and files and includes them under *Files required for your application to run*. MATLAB may not detect all files, so make sure they are all in the list. For instance, MATLAB Compiler does not seem to recognize reporting templates for the MATLAB Report Generator, so these have to either be added manually to the list, or added manually to the output after compilation. In our case, we will add the "myTemplate.pdftx" under the heading *Files installed for your end user*. Continue by selecting whether or not the MATLAB Runtime should be included in the package. Keeping the *Runtime downloaded from web* ticked will make sure the program footprint will be as small as possible. Press *Package*. The output will consist of three folders and a log file. The install file in the "for_redistribution" can be used to install the compiled app and MATLAB Runtime on a computer. Alternatively, if the appropriate version of MATLAB or the MATLAB Runtime is already installed on the computer, the folder named "for_redistribution_files_only" contains the main executable. On a Windows system, this file is called "calcSNR.exe"[4]. To run the program, simply place the DICOM image in the folder with "calcSNR.exe" and run the executable[5].

5.4 CONCLUSION

It may be beneficial to consider designing your code for deployment as an application. A stand-alone application will run independently from the creator of the code, often in automated form, and as such certain design aspects needs to be considered that are slightly different from those used during regular prototyping and scripting. This chapter demonstrated an example method designed around the relatively simple task of calculating the SNR inside a ROI placed on flat-field radiography images. In this example, some important design aspects such as identifying the correct input data, performing the data analysis, reporting of the results and storing of the results in a structured database were all demonstrated.

While this chapter focused on a rudimentary image analysis, simple report formatting and setting up a lightweight SQL database, all the concepts and code can be expanded for more sophisticated tasks.

MATLAB toolboxes used in this chapter:
Image Processing Toolbox
Database Toolbox
MATLAB Report Generator

Index of the in-built MATLAB functions used:

append	exec	mkdir	strcmp
close	exist	numel	sum
createTemplate	fetch	print	tic
delete	fullfile	rectangle	toc
dicominfo	imagesc	round	unzipTemplate
dicomread	insert	size	zeros
dirPlus	isdicom	sqlite	zipTemplate
erase	mean2	std2	

---

[4]When compilation is performed on Unix-like systems, the "for_redistribution_files_only" folder contains a binary file "calcSNR" and a shell script "run_calcSNR.sh". The latter can be executed with the path to MATLAB passed as an argument. This will configure the appropriate environment variables and then execute the "calcSNR" file.

[5]Note that *which* MATLAB Runtime version is installed is critical. If running a program without prior installation using the installer file, the user needs to ensure that the MATLAB Runtime is for the same version of MATLAB that the app was compiled on.

## CHAPTER 6

Integration with other programming languages and environments

Gavin Poludniowski

Medical Radiation Physics and Nuclear Medicine, Karolinska University Hospital, Stockholm, Sweden

Department of Clinical Science, Intervention and Technology, Karolinska Institutet, Stockholm, Sweden

Matt Whitaker

Image Owl, Inc., Greenwich, NY, USA

CONTENTS

MATLAB is undoubtedly a good choice for scientific programming. However, there is a whole other world out there of executable programs, programming languages and libraries. This chapter shows you how to take advantage of this software without leaving MATLAB behind.

6.1 INTRODUCTION

There are times when the core functionality of MATLAB and our list of toolboxes cannot conveniently solve the problem at hand. At such points, we may turn to scripts and libraries written by a third party (i.e., not ourselves or MathWorks). We may be lucky and find that somebody else has made MATLAB software available for the task. Assuming they have commented and documented their code well (see Chapter 7 on *Good Programming Practices*), it should be straightforward to integrate with our program. This chapter, however, covers the integration of software written in other programming languages and environments. The external software could include code written by ourselves as well as by others. Sometimes it is just easier to do things outside of MATLAB.

Here we will explore integrating with operating system (OS) commands, and software developed in Java, Python and the .NET framework. This list is by no means comprehensive. For example, we will not discuss MATLAB MEX files and integration with C/C++ and Fortran code. However, the examples in the chapter will give a taste of what is possible. We assume that the reader is working primarily in MATLAB and wants to access functionality in an external program or library. It is also quite possible to go the other way: integrate with MATLAB *from* external applications. We leave this for the reader to explore on their own.

Note that integration of MATLAB with external software involves complexities that are, in most cases, hidden from the user. The reader must be prepared to do some detective work when things go wrong. We will try to be explicit regarding the distributions, releases and operating systems with which we obtained the various results, to minimize any difficulties. MATLAB R2019b was used to develop the examples and they should work with later versions without issue.

6.2 WHEN TO USE OTHER PROGRAMMING LANGUAGES AND ENVIRONMENTS

In deciding to work with MATLAB, the user is choosing a tightly bundled package of tools, optimized and improved over decades. MathWorks provides a platform-independent execution engine that runs your MATLAB code, a single desktop environment to develop your code and an extensive library of well-documented functions to assist you in solving your programming problems. The user experience is also quite consistent across multiple operating systems and MATLAB code is largely independent of the user's OS. Typically the user has few choices to make, excepting which license to select and which toolboxes to purchase. In the case of other programming languages, this is often not the case. There may be a choice of several compilers or runtime environments, various available development environments for coding and a host of available libraries, possibly outside of centralized control. While this breadth of choice undoubtedly provides excellent opportunities, the sheer number of *options* can be overwhelming for the novice.

It may be that you have only ever used MATLAB, that you think it logical and efficient, and you have never seen the need to integrate it with another platform. Fine. Remember, though, MATLAB may only seem best for all your tasks because you are most familiar with it. And when the only tool you have is a hammer, everything begins to look like a nail. Yet MATLAB *is* more of a Swiss Army Knife than a hammer. Depending on your tasks, you might be able to get by happily with it alone. If so, good luck to you. If not, read on.

The are many reasons why you might want to integrate with software from another programming platform, but no clear rules for when you should. Some situations you might encounter are:

- The functionality I want already exists in a library for another programming language and I do not want to re-invent the wheel;

- MATLAB cannot do what I want, or I cannot get it to do what I want;

- I am collaborating with colleagues who are working in another programming language for their part of the project;

- I like another programming language for performing a specific task, but I do not want to lose the power and convenience of MATLAB;

- My line-manager or supervisor told me to do it.

In some cases, external software may readily solve your task for you. Whether you *should* use it is another matter. It could depend on the conditions on which it is made available (see the discussion of licensing in Chapter 8), and how well-validated it is (see the topic of off-the-shelf software in Chapter 9).

For the examples explored here (in Java, Python, .NET), it is straightforward to get started working from the MATLAB desktop environment. Note that Java is bundled with MATLAB and MATLAB will use your computer's .NET version. For Python, you will have to install a version of the standard implementation (also known as CPython) if not installed already, but this should not provide great difficulties. First, however, we will explore the execution of system commands from MATLAB.

6.3   SYSTEM COMMANDS

In this context, the system means your *operating system*. We will assume you are running MATLAB on a Microsoft Windows or a UNIX-like system (tip: Apple's Mac OS is UNIX-like, as are Linux and BSD). It can be useful on occasion to use built-in commands of your operating system. You can do this using the `system()` function of MATLAB. This function tells the operating system to execute a command specified by a string or character array supplied as an argument. For example, on Windows, to list the contents of the current folder, you can do the following:

```
[status, cmdout] = system('dir');
```

The function starts a new Windows *cmd* process and executes the specified instruction. The operation waits for the command to finish before returning the exit status and command output. The command returns a number in `status` (0 for a successful call) and any output in `cmdout` (a character array). On a UNIX-like system, you would instead do the following:

```
[status, cmdout] = system('ls');
```

where the command is executed in a new shell process. Note that even though the MATLAB call is platform-independent, the command passed to the OS is not.

The above example was about as simple as it gets. MATLAB actually has its own command to do precisely the same task (called `ls`). However, it is not hard to think of slightly more advanced operations where it makes sense to leverage an operating system's capabilities, for example, listing all running processes (use the `tasklist` command for this in Windows). The `system` command can also run any executable script or program. We will explore this further.

In Chapter 5, an example image analysis task using an image-viewing program was discussed. The workflow described was manual and not easily exactly repeatable, involving region-of-interest (ROI) placement by hand. Some programs have solutions for such issues. The ImageJ application (National Institutes of Health, USA) has a powerful macro capability for the scripting of actions, commands and calculations. We will be using a distribution of ImageJ2 called Fiji, which includes many plugins for image analysis[1]. As an example, we have created an ImageJ macro for the task of opening up a DICOM image, placing an ROI and extracting the mean and standard deviation. ImageJ can automatically generate such a macro by capturing your actions in the GUI (under *Plugins-Macros-Record* in the menu). The resulting macro "roiMacro.ijm" is presented below. Remember that this is *not* MATLAB syntax.

---

[1]Find the Fiji distribution here: `https://fiji.sc`. The installation of Fiji used for this chapter was based on ImageJ 2.0.0-rc-69/1.52p.

```
//ImageJ macro
run("Set Measurements...", "mean standard redirect=None decimal=3");
open("myProjXray.dcm");
makeRectangle(960, 960, 100, 100);
run("Measure");
```

The **run** command in the macro is used to execute options from the ImageJ toolbar (*Analyze-Set Measurements* and *Analyze-Measure* in this case). The **open** command opens up an image file. Here, we make use of the flat-field planar image shown in Figure 5.4 in the previous chapter. The **makeRectangle** command places a rectangular ROI; the first two arguments define the upper-left corner in pixels and the second two the size. The macro assumes a known image size of 2022×2022 pixels. The image file is supplied with this book, but before running this macro, the reader must download the image file and make sure ImageJ can find it. To do the latter, edit the macro to specify the path to the image file.

We will now step through some MATLAB code to calculate signal-to-noise ratio (SNR). We assume that your computer's system path includes the ImageJ executable (e.g., "ImageJ.exe") and that the macro and DICOM image are placed in MATLAB's working directory. Then the macro can be run in ImageJ from MATLAB as follows:

```
% Replace "ImageJ.exe" with other binary name if appropriate e.g. ImageJ-linux64
command = 'ImageJ.exe --headless -macro roiMacro.ijm';
[status, cmdout] = system(command);
```

The **--headless** flag instructs ImageJ to not display its graphical interface and **-macro** tells ImageJ to execute a macro file[2]. The work does not stop here, however. The results have been returned in **cmdout**, but we have to extract them from the character array. One way (of many) is as follows:

```
cmdoutSplit = splitlines(cmdout); % Split the output by lines
matchArray = contains(cmdoutSplit,'StdDev'); % Get binary array for string matches
iMatch = find(matchArray); % Find index of the non-zero element (the matching line)
dataLine = char(cmdoutSplit(iMatch+1)); % Get the next line and convert from cell
dataLineSplit = strsplit(dataLine); % Split character array by white spaces
meanVal = str2double(dataLineSplit(2)); % Extract mean, convert to double
stdVal = str2double(dataLineSplit(3)); % Extract stdev, convert to double
```

As you can see, extracting the critical information from **cmdout** took more lines than generating the data in the first place. It is not uncommon that working out how to convert data inputs and outputs to the appropriate form is time consuming and sometimes it is the most challenging part of interfacing with external software.

Note that an alternative approach, in this case, would have been to save the output of ImageJ to a text file and then read it into MATLAB. Regardless of our choice, we can now finish with calculating the pixel-wise SNR. This can be obtained in MATLAB as follows:

```
SNR = meanVal/stdVal
```

For the test image, this produces the output:

```
  SNR =

      117.5793
```

which is consistent with the results of Chapter 5.

---

[2]The precise flags required can vary with version and release of ImageJ (in particular for ImageJ1). You may need to replace "–headless -macro" with an alternative if you are not using the Fiji distribution.

6.4 INTEGRATING WITH JAVA

Java is a general-purpose programming language first released in the 1990s. Although a private company owns Java (Oracle Corporation), there are open-source versions and, at the time of writing, it is free to obtain and use for non-commercial purposes. Java has become very popular and is currently running on billions of devices worldwide, in a wide variety of applications and environments. Programs written in Java are very portable (can run on many different computer architectures and operating systems) and there is a vast body of applications and libraries available to help the software developer. Note that Java is not JavaScript, which is an entirely distinct programming language.

Java is known for its C/C++ like syntax and like those programming languages, it is a compiled language. Also, like C/C++, it is statically typed, rather than dynamically typed. Unlike with MATLAB therefore, the programmer must declare in code what data type a variable is before compilation, rather than allowing the interpreter to infer data type at runtime. In Java, however, programs are not compiled into executable files, as they are in C/C++. Instead, they are compiled into something called *bytecode*, which is then executed by the Java Virtual Machine (JVM) at runtime. The bytecode is independent of the machine on which it is run, which contributes to making it very portable. Java also pioneered the idea of the just-in-time (JIT) compiler—something that MATLAB now also has—the use of which means that code is often fast to execute. Java is not, however, interpreted, in the same sense MATLAB is. The concept of an interpreter that you can type commands into is not central. Java programs package all code into classes and the language is fundamentally object-oriented in approach (see Chapter 2, *MATLAB fundamentals*, for an introduction to classes).

You can check what JVM version MATLAB is using on your computer by typing:

version -java

For our installation (MATLAB R2019b), the command output informs us that we have "Java 1.8.0_202-b08" installed, which is a build of Java 8.

To make use of in-built Java libraries in MATLAB, type **java** followed by the namespace and method of interest. The following returns a list of files and directories in your current working directory.

```
java.io.File('.').listFiles()
```

Here, **io** is the Java package, **File** a class in that package and **listFiles()** is a method in that class. The line of code returns an array of filenames, or more precisely a *Java array* object containing *Java File* objects. This can be converted to MATLAB types, for example, to a string array using MATLAB's **string()** function. Note that you can check which methods belong to any Java class using MATLAB's **methods(**classname**)** or **methodsview(**classname**)** function.

To use external Java classes (.class files) or libraries (typically .jar files containg classes), MATLAB has to be able to find them. MATLAB uses the *Java class path* to search. This is split into *static* and *dynamic* parts. You can display the locations on the path using the **javaclasspath** command. To add compiled Java files so that they are persistent across MATLAB sessions, you can create a "javaclasspath.txt" file in your preferences directory (type **prefdir** to find this) and add the locations to it. This adds them to your static path. Alternatively, you can temporarily add them to your dynamic path using the **javaaddpath** command. See the MATLAB documentation for further details.

For an in-depth example of combining Java with MATLAB, we will explore interfacing with ImageJ. We saw an indirect way of doing this in the previous section, using the **system** function. While this works and is the simplest approach in some cases, it is challenging to leverage the full power of ImageJ in this fashion. Fortunately, others have encountered this issue previously and come up with solutions, making use of the fact that ImageJ is a Java application and that Java is integral to MATLAB. As in the previous subsection, we will use a distribution of ImageJ called Fiji. We will assume that you have downloaded and installed it. Once you have opened up Fiji, you will need to select *Help-Update* and add the *ImageJ-MATLAB* extension. Once you have installed this you are ready to get started. We will begin within the MATLAB interpreter. Type (editing the path as appropriate):

```
addpath '/path/to/Fiji.app/scripts' % Customize this line for your system
```

This allows MATLAB to find the Fuji's scripts. Now type the next two lines:

```
ImageJ;
ij.IJ.run('Quit','')
```

The first line runs a MATLAB script called **ImageJ.m** in the **scripts** folder. This loads the ImageJ libraries, making use of the MATLAB function **javaaddpath** to add the necessary Java archives (.jar files) to the dynamic path. It also starts an instance of the ImageJ toolbar. The second line *closes* that toolbar. We will not need it, as we will be working entirely in the interpreter. Note that **ij** is the main ImageJ Java package, **IJ** is a class in the package and **run** is a method in the class.

Below is a short code snippet that performs the same task as the macro discussed in the previous section. As before, you must ensure that ImageJ (Fiji) can find the file "myProjXray.dcm", specifying the path if it is not in your working directory.

```
imp = ij.IJ.openImage('myProjXray.dcm'); % Read in an image (DICOM) file
impDim = imp.getDimensions(); % Get the dimensions of the image
width = impDim(1);
height = impDim(2);
imp.setRoi(width/2-51, height/2-51, 100, 100); % Put ROI at centre of image
stats = imp.getStatistics(); % Get statistics for the ROI
imp.deleteRoi(); % Delete ROI (not necessary here, but can be useful)
meanVal = stats.mean; % Get mean pixel-value in ROI
stdVal = stats.stdDev; % Get standard deviation in ROI
SNR = meanVal/stdVal % Calculate pixel SNR
```

As in the previous section, with the same image, this produces the output:

```
  SNR =

      117.5793
```

As can be seen, this is an elegant approach. Here the analysis was easily generalized to an image of arbitrary size, using the **getDimensions** method of the **ImagePlus** class, of which the **imp** object is an instance. However, with the ImageJ-MATLAB extension, we can do much more than re-implement the job done by a macro. We will now work directly in the interpreter again, typing in commands:

```
imp = ij.IJ.openImage('mySliceCT.dcm');
```

This line reads in a new DICOM image (a CT slice of a phantom this time). Another ImageJ Java class worth knowing about is the **DicomTools** class in the **ij.util** package. We can find what methods it contains by using the **methods** function:

```
methods('ij.util.DicomTools')
```

This produces the output:

```
  Methods for class ij.util.DicomTools:

    DicomTools    getTag        hashCode
    equals        getTagName    notify        toString
    getClass      getVoxelDepth notifyAll     wait
```

We are interested in the `getTag` method. We can use it to extract the study date for the DICOM image by passing the DICOM tag as an argument:

```
studyDate = ij.util.DicomTools.getTag(imp,'0008,0020')
```

This gives:

```
  studyDate =

    20180201
```

We will now show the image slice using an ImageJ window. This can be achieved as follows:

```
imp.show()
```

The image is illustrated in Figure 6.1a. We can get rid of the image window as follows:

```
imp.hide()
```

In the following four lines we will copy the image, run the *Process-Binary-Make Binary* and *Process-Find Edges* options of ImageJ on the copy and show the resulting image:

```
imp2 = imp.duplicate();
ij.IJ.run(imp2, 'Make Binary', '');
ij.IJ.run(imp2, 'Find Edges', '');
imp2.show() % Show image with ImageJ
```

The processed image is shown in Figure 6.1b.

This was interesting. However, we would like to be able to transfer image data conveniently between ImageJ and MATLAB. The ImageJ-MATLAB extension provides some tools to do this. In your MATLAB workspace, you should find a variable called `IJM` (a Java object). `IJM` has several useful methods, including `IJM.show(arrayName)` and `IJM.getDatasetAs(arrayName)`. The former takes a named MATLAB array and displays it in an ImageJ window. The latter takes the data from the active ImageJ window and puts it in a MATLAB array with the specified name. In the three lines below, we will see image data extracted, the ImageJ window hidden and the image data displayed using the native MATLAB function `imshow`.

```
IJM.getDatasetAs('myArray') % Put the data in a MATLAB array
imp2.hide() % Close the image in ImageJ
imshow(myArray') % Display the image using MATLAB
```

The resulting image is displayed in Figure 6.1c. Note that the transpose of the array was passed to `imshow`, in order to flip the axes and make sure the image was displayed correctly.

For more on the ImageJ-MATLAB extension, see the ImageJ website[3]. For another example of using Java in MATLAB (the EtherJ DICOM library), see Chapter 18 (*Image processing at scale by "containerizing" MATLAB*).

6.5 INTEGRATING WITH PYTHON

Python, like Java, is a general-purpose and portable programming language. It was also first released in the 1990s. Its development is managed by the Python Software Foundation, which is committed to it remaining open-source and completely free. Python's design philosophy emphasizes conciseness and readability of code. It is famously simple to learn and has a vast array of libraries available.

As with Java, Python code compiles to bytecode, and then executes on a virtual machine (a Python Virtual Machine in this case). An important aspect of Python, however, is that

---

[3]Details on interfacing ImageJ and MATLAB can be found here: `https://imagej.net/MATLAB_Scripting`.

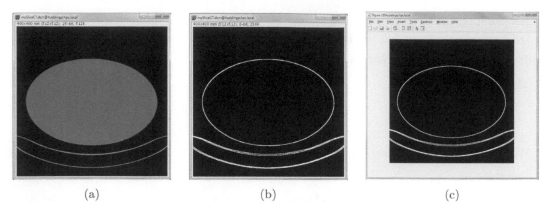

(a)   (b)   (c)

Figure 6.1: CT slice of a phantom displayed in MATLAB. Images (a) and (b) are displayed using ImageJ and depict the slice before and after running the ImageJ *Make Binary* and *Find Edges* options, respectively. Image (c) presents the image data using the `imshow` function of MATLAB.

this compilation is implicit. The user never invokes a compiler but simply runs Python code. Like MATLAB, Python is dynamically-typed, which means that the programmer does not need to explicitly specify the type of variables and can let the virtual machine infer them at runtime. Also like MATLAB, and unlike Java, Python features an interactive interpreter. The programmer can type Python statements in and have them executed immediately. Like in Java and MATLAB, object-oriented programming can be adopted in Python (and we will see an example of a class later); however, as with MATLAB, this is not obligatory.

Two standard libraries that are essential for most scientific programming in Python are *NumPy* and *SciPy*. The *PyDicom* library is also handy for processing medical images. The *matplotlib* library can be used for plotting data. Beyond these core packages, there are frequently alternative libraries available to do specific tasks, although the documentation and details of validation may be sparse.

MATLAB does not include Python in its installation and so to begin using it you will need to make sure you have a suitable version installed on your computer. MATLAB (R2019b) officially supports versions 2.7, 3.6 and 3.7 of the reference implementation (CPython) available free-of-charge from the Python Software Foundation[4]. Anaconda Python (Anaconda, Inc.) is a popular free alternative that bundles many useful libraries. In our experience, this also works fine with MATLAB[5].

Once Python has been installed, you will need to make sure MATLAB can *find* it. If the system search paths are configured so that you can invoke Python from a console window (e.g., from *cmd.exe* on Windows or a *terminal* in Linux), then MATLAB should find your Python distribution. To check, type `pyenv` (or `pyversion` if your MATLAB version is older than 2019b)[6]. Assuming everything is in order, you can make use of Python libraries by

---

[4] Available for download here: `www.python.org`.

[5] The code of this section has been successfully tested with various combinations of the following: Python 3.7 and 3.8; CPython and Anaconda distributions; Windows, Linux and Mac OS operating systems. Python2 is no longer under development, so we recommend the reader uses a Python3 version rather than Python 2.7.

[6] If MATLAB cannot find your Python distribution, you can manually add the path with: `pyenv('Version',executable)`, where `executable` is the path to your Python executable. If the *library* entry of the output from `pyenv` is empty, then your Python distribution is probably compiled without shared libraries enabled. In this case you will either have to install an alternative distribution or re-compile Python from source with the "–enable-shared" option selected.

typing **py**. It is essential to know that MATLAB loads the Python interpreter at the first use of **py** and it is not possible to switch Python versions afterwards in the session. The following line of MATLAB code uses the `listdir()` function of the **os** Python module to list the files and directories within your current working directory.

```
py.os.listdir('.')
```

The output of the above command is a list. To be more specific, it is a *Python list* of *Python strings*. The list can be converted to MATLAB types, for example, to a string array, using MATLAB's `cell()` function followed by its `string()` function.

The conversion back-and-forth between MATLAB and Python data types is possibly the trickiest aspect of using Python in MATLAB. In some cases, this can be done automatically by MATLAB. For example, the MATLAB character array `'.'` was provided as an argument to the Python function (telling it to look in the current directory) and this was converted to a Python string automatically. In many cases, however, it is up to the programmer to sort out type conversion.

For a non-trivial illustration of integration with Python, we will make use of the *SpekPy* package. SpekPy is a toolkit for calculating x-ray spectra from x-ray tubes and can be downloaded free-of-charge[7][19]. We assume you have successfully installed SpekPy on your computer before proceeding, following the instructions in SpekPy's homepage. We will walk through a code section that calculates and plots a spectrum. First, we want to create a SpekPy spectrum model (an instance of a Python class called **Spek**). In pure Python, we would do it like this: `s = spekpy.Spek(kvp=80,th=12)`. In the expression, there are two keyword arguments: **kvp** (x-ray tube potential in kV) and **th** (the anode angle in degrees). A keyword argument in Python is simply a named argument to a function. MATLAB, however, does not pass arguments to functions in the same manner. When the Python programmer would use keyword arguments in their function signatures, the MATLAB programmer needs to use the **pyargs** function to prepare the input. We create the spectrum in MATLAB as follows[8]:

```
pot = 80; % The x-ray tube potential in kV
ang = 12; % The target anode angle
spekpyArgs = pyargs('kvp', pot, 'th', ang); % Define Python keyword args
s = py.spekpy.Spek(spekpyArgs); % Create a spectrum model with arguments
```

We can add filtration as follows:

```
s.filter('Al',6.0); % Filter the spectrum with 6 mm of aluminium
```

Now it is possible to calculate useful quantities associated with the spectrum. One of these is the first half-value-layer (HVL) in mm of aluminium:

```
hvl = s.get_hvl1()
```

This gives the output:

```
  hvl =

    4.3750
```

---

[7]Find the SpekPy package here: `https://bitbucket.org/spekpy/spekpy_release`. SpekPy Version 2.0.0 was used for this chapter.

[8]With Linux installations of Python, we have encountered some compatibility issues with NumPy/SciPy and MATLAB's use of the Intel MKL libraries. If SpekPy causes MATLAB to crash, try executing the following MATLAB statement before calling SpekPy: `py.sys.setdlopenflags(int32(bitor(2,8)))`. This should solve the problem (see: `https://mathworks.com/matlabcentral/answers/358233-matlab-python-interface-broken`).

In this case, MATLAB automatically converts the Python output data type (`numpy.float`) to a MATLAB equivalent (`double`). You can verify this using the expression `isa(hvl,'double')`, which will output a value of 1 (i.e., the assertion is true).

We can also extract and plot the spectrum itself.

```
spectrumArgs = pyargs('edges',true); % Define Python arguments
spectrumData = s.get_spectrum(spectrumArgs); % Get the spectrum data
```

The output is in histogram bin (staircase) format because the `edges` keyword was set to *true*. It was simple to obtain the spectrum, but to use it requires more work. MATLAB does not handle the type conversion here and we must deal with it ourselves.

```
kNumPy = spectrumData{1}; % Extract the energy bins
fNumPy = spectrumData{2}; % Extract the fluence values
kListPy = py.list(kNumPy); % Convert NumPy array to Python list
fListPy = py.list(fNumPy); % Convert NumPy array to Python list
kCellMAT = cell(kListPy); % Convert Python list to MATLAB cell array
fCellMAT = cell(fListPy); % Convert Python list to MATLAB cell array
k = cell2mat(kCellMAT); % Convert MATLAB cell array to double array
f = cell2mat(fCellMAT); % Convert MATLAB cell array to double array
```

This conversion was a cumbersome process. If you have to convert from one type to another frequently, using multiple steps, it is worth considering writing a function to do the task.

With type conversion out of the way, we can finally plot our data.

```
plot(k, f) % Plot spectrum
xlim([0 90])
xlabel('Energy bin  [keV]')
ylabel('Fluence per unit energy @ 1 m  [cm^{-2} keV^{-1}]')
title('X-ray spectrum')
```

The plot should look like Figure 6.2. For a library with comparable capabilities to SpekPy but written in MATLAB, see Chapter 16 (*xrTk: MATLAB a toolkit for x-ray physics calculations*).

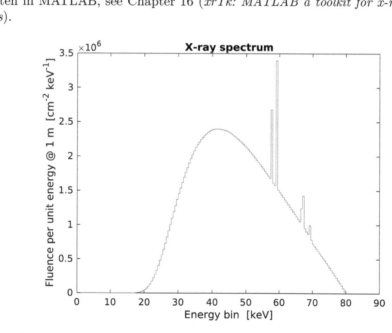

Figure 6.2: An x-ray spectrum as calculated by the SpekPy Python toolkit and plotted in MATLAB.

6.6 INTEGRATING WITH THE .NET FRAMEWORK

Fewer people are aware of MATLAB's extensive support for the .NET framework, compared to that of, for example, Java libraries. The .NET framework, developed by Microsoft, employs a comparable model to Java in that compiled code is run on a virtual machine called the Common Language Runtime (CLR) similar in some respects to Java's Virtual Machine.

Although primarily encountered in Windows environments, the .NET system is technically platform-independent. Microsoft has made .NET Core libraries open-source and it has appeared in other environments such as Linux[9]. One of the advantages of using the .NET platform is it gives access to a wide range of stable and well-documented libraries that make dealing with the Windows environment convenient. Like Java and Python, there is an active developer network producing a wide range of useful libraries for the Medical Physics community. At the time of writing (R2019b), the .NET MATLAB interface requires a Windows installation and .NET Framework Version 4.0 or above[10].

Libraries of classes in .NET are termed assemblies and generally have the .dll extension. The .NET Global Assembly Cache (GAC) registers most of the .NET core libraries. It is beyond the scope of this discussion, but developers can add their assemblies to the GAC for easy sharing. For libraries that are not part of .NET's core, users will generally reference what are termed private assemblies by supplying the full path and name of the DLL file when loading the assembly.

MATLAB includes a suite of functions to support .NET integration under the NET namespace. To see an overview of functions in the .NET interface, type: **doc NET**. Before attempting to use .NET, use the **NET.isNETSupported** function to confirm that a supported .NET framework is available. In most modern Windows installations this is not an issue.

The first time we use a NET interface function or a function within the .NET *System* library, MATLAB loads specific core .NET libraries. For .NET libraries within the **System** namespace, it is often not necessary to explicitly load the assemblies. For example, to get the filenames of all files in the current directory, you can use the **System.Directory** library:

```
System.IO.Directory.GetFiles('.','*',System.IO.SearchOption.TopDirectoryOnly)
```

This outputs the names of all files in the current directory. To be more precise, it returns a .NET object, containing a set of *.NET string* objects. This can be converted to MATLAB types, for example, to a string array, using MATLAB's **string()** function.

For most assemblies, it is necessary to explicitly add the assembly to make it visible in MATLAB. For assemblies registered in the GAC, you can load the assembly using its namespace. The following code uses the **System.Speech** assembly to vocalize some input text through your computer's speakers.

```
asm = NET.addAssembly('System.Speech');
speechSyn = System.Speech.Synthesis.SpeechSynthesizer;
speechSyn.Volume = 70;
Speak(speechSyn,'I am using .NET')
```

Note that the **NET.addAssembly** function returns an instance of a useful class called **NET.Assembly** that contains details on many aspects of the added library.

If the assembly is private and not registered, it can be added by passing the fully qualified name of the assembly DLL. We see this in our more in-depth .NET example, using the Evil DICOM library[11]. Evil DICOM is an example of the many useful third-party open-source

---

[9]In fact, Microsoft has announced that in future releases, the primary implementation of .NET (where the "framework" will be dropped from the name) will be based on .NET Core. See, for example: https://devblogs.microsoft.com/dotnet/introducing-net-5. MathWorks's timeplan for implementing support for this is unclear at the time of writing.

[10].NET assemblies built on Framework 2.0 and above can be loaded and used.

[11]See here for more on Evil DICOM: https://github.com/rexcardan/Evil-DICOM.

libraries available in the .NET ecosystem. For users who do not have access to MATLAB's *Image Processing Toolbox* for DICOM manipulation, Evil DICOM (written in the .NET language of C#) provides simple tools for examination and manipulation of DICOM files. The library also contains some essential DICOM network tools not included with the *Image Processing Toolbox*. You can download the package in Microsoft's *nuget* format. If you have admin rights on your computer, you can install following the command-line instructions listed on the download page[12].

The example below shows the contents of a code section with simple examples of finding tags in DICOM files. We also demonstrate a simple tag modification and saving the modified data to a new DICOM file. We assume that you have installed the Evil DICOM DLL file.

First, we need to add the assembly.

```
evilDicomAssembly = 'EvilDICOM.dll'; % Name of assembly
pathToEDDll = '/path/to'; % Path to assembly (edit as appropriate)
NET.addAssembly(fullfile(pathToEDDll,evilDicomAssembly)); % Add assembly
```

Then we need to import the namespaces we wish to use from the Evil DICOM library.

```
import EvilDICOM.Core.*;
import EvilDICOM.Core.Helpers.*;
import EvilDICOM.Core.Selection.*;
import EvilDICOM.Core.Enums.*;
import EvilDICOM.Core.IO.Writing.*;
```

Now we can load a DICOM file into MATLAB using the library. We will use one distributed with MATLAB.

```
dcmFile ='CT-MONO2-16-ankle.dcm';
dcm = DICOMObject.Read(dcmFile); % Read dicom file into a DICOMObject
```

Next we can begin to extract information, such as the `PatientName` tag.

```
patientName = dcm.FindFirst(TagHelper.PatientName); % Find patient name tag
```

The DICOM dictionary functionality built into Evil DICOM allows for sophisticated searches and intelligent navigation of the tags. For example, the following code identifies all elements that are of Value Representation (VR) *Person Name* type and prints the attribute names.

```
allPersonNameElements = dcm.FindAll(Enums.VR.PersonName); % A .NET List
for n = 0:allPersonNameElements.Count-1 % List is indexed from 0 (not 1)
    thisElement = allPersonNameElements.Item(n);
    % Lookup the description for each returned element
    desc=char(EvilDICOM.Core.Dictionaries.TagDictionary.GetDescription(...
        thisElement.Tag.CompleteID));
    disp(['Tag name: ' desc]); % Display each description to screen
end
```

The `DICOMSelector` object is convenient for quick selection and manipulation. The following allows us to get the image size quickly.

```
sel = dcm.GetSelector();
rowSize = sel.Rows.Data;
columnSize = sel.Columns.Data;
```

---

[12]To download Evil DICOM, see: https://www.nuget.org/packages/EvilDICOM. If you do not have admin rights, you can download the nuget package, change the suffix from *nupkg* to *zip* and then unzip it. The DLLs can be found therein and an appropriate one can simply be loaded using the MATLAB `addAssembly()` function. If this installation method is adopted, we recommend the reader work backwards through the releases to find a version with no dependency requirements (we used Evil DICOM 2.0.5.7 on both Windows 7 and Windows 10).

It is also possible to modify and save data. In the next snippet of code, we modify the patient name, save the modified DICOM file under a new filename, then read it back in and confirm the value is changed correctly.

```
originalValue = char(sel.PatientName.Data);
newValue = 'Modified using Evil DICOM';
sel.PatientName.Data = System.String(newValue); % Modify PatientName string
sel.ToDICOMObject.Write('ModifiedSample.dcm'); % Write modified DICOM file
dcmMod = DICOMObject.Read('ModifiedSample.dcm'); % Read it back in
selMod = dcmMod.GetSelector();
disp(['Original patient name: ' originalValue]); % Display
disp(['Modified patient name: ' char(selMod.PatientName.Data)]); % Display
```

The output to the MATLAB command window from running the entire Evil DICOM code section is:

```
Tag name: ReferringPhysicianName
Tag name: NameOfPhysiciansReadingStudy
Tag name: OperatorsName
Tag name: PatientName
Original patient name: Anonymized
Modified patient name: Modified using Evil DICOM
```

This example only begins to show what is possible with this powerful DICOM library.

6.7 CONCLUSION

This chapter has given a sampler of what is possible when you begin to make use of software developed outside of the MATLAB environment. It is not a case of either one programming language or another. In many cases, it is possible to integrate with MATLAB allowing us access to a richer world of libraries and resources.

The key steps to successfully using external packages with MATLAB are:

1. Inspect the appropriate MATLAB documentation;

2. Ensure that you have a compatible version of the external software environment installed along with the package of interest;

3. Check that MATLAB can find it, for example, using your computer's system path or the addpath, javaaddpath, pyenv and NET.addAssembly commands in MATLAB;

4. Load the library with the correct syntax, as specified in the MATLAB documentation.

MATLAB toolboxes used in this chapter:
Image Processing Toolbox (for example, DICOM data sets)

Index of the in-built MATLAB functions used:

addpath	imshow	pyenv	title
cell	javaaddpath	pyversion	version
cell2mat	javaclasspath	splitlines	xlabel
char	ls	str2double	xlim
contains	methods	string	ylabel
disp	methodsview	strip	
find	plot	strsplit	
fullfile	pyargs	system	

## CHAPTER 7

Good programming practices

Yanlu Wang

Department of Clinical Sciences, Intervention and Technology, Karolinska Institutet, Stockholm, Sweden
Medical Radiation Physics and Nuclear Medicine, Karolinska University Hospital, Stockholm, Sweden

Piyush Khopkar

MathWorks Inc., Natick, MA, USA

CONTENTS

L EARNING to program is more than being able to write functioning code. Depending on who you ask, the definition varies greatly on what is expected from good programming. In this chapter, we focus on what one might expect from good programming practice in terms of the code being produced. In line with what my mentor once told me: "There are many ways to skin a cat". However, as we all know thanks to George Orwell, "All animals are equal, but some are more equal than others". The same thing can be said about writing code.

7.1 WHAT MAKES A GOOD PROGRAM

First and foremost, a good program works. No program can be considered good when it cannot even fulfill its purpose. A program's ability to function as intended is a necessary condition, but often not sufficient. Here are four important aspects to consider:

Efficiency. A good program should be efficient when performing its task. Efficiency is often synonymous, though not always equivalent, to fast. For the sake of brevity, we simply state that: an efficient program processes its task without occupying more resources than needed in a reasonable amount of time.

Simplicity. A good program should be simple and intuitive to use. As with most of these points, it is difficult to specify details on what this entails in practice, as different programs have varying intended purposes and users.

Flexibility. A good program should be flexible, within reason. General purpose programs by their very nature tend to be large and difficult to maintain. A good program should be designed to perform a specific task, but should also be flexible enough to perform variations of the task. What constitutes variations of a task, and another task altogether, is up to the author to decide.

Well-structured. Not unrelated to the previous point, a good program should also be well-structured programmatically. It is in your own best interest to structure the program to accommodate future changes and facilitate extension of the program's functionality. Having well-structured code also eliminates a lot of the overhead involved when collaborating with other programmers. There is nothing worse for a programmer than having to read badly structured code.

7.2 GOOD PRACTICES

In the remainder of the chapter we briefly mention some tips and tricks to help turn you from a person who knows how to code, into someone who writes good programs. Many of these tips may be obvious to the experienced programmer, and to those who tend to read about programming more than they do it, but do not underestimate these very pedestrian warnings! We promise you the reader, the effort spent on these tasks will be worth it. The space allocated for these topics in this book does not do them justice, as each could fill a book in its own right. Below are simply brief introductions to each topic, and some basic pointers to get you started.

7.2.1 Commenting

A comment in MATLAB begins with the % symbol. Nothing on a line after this symbol will be executed as code. Simply saying "comment your code" would be lazy, because that much is universally true. There are many reasons to why you should do this, and we assume that you have some of these figured out by now. If not, then nothing we write here will make you do it anyway. But for the sake of completeness, we will give you two reasons and some rules of thumb along the way: comment for yourself, and comment for others. The length and frequency of the comments in each category depends on you and who exactly you are writing the code for.

Comment for yourself
Your "Eureka!" moments are called that precisely for a reason: You do not have them all the time. It is guaranteed that you will forget why you wrote your code that way, and you will forget why it worked.Do not let your moments of insight go to waste. At the very least, write to your future, more stupid self, why you are currently so smart. If any line of code is not trivially obvious when you are writing it, it probably will become much worse tomorrow, and a mystery in a week. Save future you from yourself and explain yourself in the present. Comments of this type are rather informal, so use whichever format you prefer and as long as necessary depending on how stupid you estimate your future self to be.

Comment for others
These are for other poor souls, who for whatever reason are forced to read the results of

your late-night, caffeine-fueled taps on the keyboard. One may think this is more or less good will, but you know hand reading other people's bad code can be maddening, so why be a part of the problem when you can be a part of the solution?

Comment whenever there is a jump in your code logic. Functions should always be commented with their purpose, and if possible, where they are used. Much like commenting for yourself, explain all non-trivial logic. Unlike commenting for yourself, nobody likes to read superfluous text in addition to a tangle of spaghetti code. Try to keep your comments concise and on point.

7.2.2 Documentation

Similar arguments apply to those in the commenting section above. There is a reason why so many frameworks exist to automatically extract comments from your code and format them into proper documentation using HTML or LaTeX.

Documentation may be aimed more towards end-users. At the very least, explain in text: (1) what your program does, (2) how to install it and (3) basic usage instructions. How to do all those things may seem trivial to you, the coder, but who else is going to know it if you do not write it down? The internet is not always a friendly place, and nobody is going risk "corrupting" their system with your code unless they know they can make use of it.

Most version control systems have mandatory documentation built in. Use this feature wisely. A tip when writing documentation for the end-user: no matter how you choose to record the information, each of the three points mentioned previously should be clear to the reader within 20 seconds. Few people will spend more than this trying to figure out the purpose of a program.

7.2.3 Version control

We will not stress the importance of keeping back-ups on your data, because that is not the focus of this book. However, version control is much like keeping back-ups, but specifically for your code! The following scenario happens to the best of us: you override your own code with better functions, or simply trim unused portion of the code because it is no longer being used. Months or even years later, someone asks specifically for a feature to be added in your program. You frown, because you remember you had that exact feature once upon a time, and remember how swiftly it got deleted; the only thing you do not remember is how you implemented it.

Here is where version control systems come in. They are programs specifically designed to keep continuous track of all changes in your code. Keeping different versions of your code in folders named after dates stopped being version control right around the time we as a species stopped worrying about the "y2k" bug. With serious constraints in space, we attempt to introduce you the reader to version-control systems: what they are, basic concepts common to them and how to use them.

Version control Systems: Git and SVN
The two most popular version control systems are Git, and Subversion (SVN). The main difference between SVN and Git is that SVN is a centralized system while Git is a distributed system. This means that SVN operates from a central server while Git keeps full copies of the code on each and everyone's computer. This two different approaches are illustrated in Figure 7.1.

Both have their up- and down-sides, and both have a healthy pedigree behind them. SVN was created in 2000 by CollabNet Inc. and is currently a top-level project of the Apache Software Foundation. SVN has a strong pedigree of projects, such as FreeBSD,

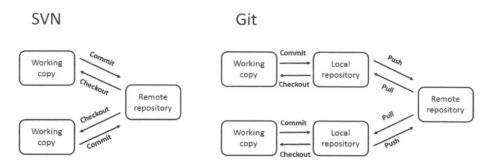

Figure 7.1: The centralized and distributed approaches of SVN and Git to version control.

GCC and SourceForge. As a centralized system, SVN allows for extensive code-base administration from a top-down perspective. Git is a distributed version control system first released in 2005. Git is by comparison, extremely fast, robust and scales well. We will not mention a long list of well-known projects that use Git, but simply state that Linus Torvalds, the original creator of Git, developed it to track developments on the Linux Kernel. Due to limited space, we will use Git as an example for the basic concepts of version control.

Version Control Systems: Basics (Git)

Here we outline the fundamental principles of Git and its terminology, most of which is shared with other version control systems. This is not an extensive overview, as there are better books for that purpose. There is simply no reason not to read the *Pro Git* book by Scott Chacon and Ben Straub, because it is free and available online[1]. Consider this a concise introduction to git for those who are new to version control systems and who wish to quickly come up to speed. For the sake of brevity, we will bull through the most essential terminology and concepts in one go. It might catch you off guard if you are yet unfamiliar with either. To help, all git terms are marked in **bold**. Do not be afraid to read this section multiple times.

Git organizes and manages projects in **repositories**: one **repository** for each project. A **repository** is "everything" in a single (root) folder. While nothing outside this root folder can be a part of the **repository**, not all files in the root folder need be **tracked**. You can **add**, or **rm** (remove), the files you wish to be **tracked** manually from the Git command line, and all other files will not be included for version control. You can **config** (configure) each and all **repositories** to automatically include files, or exclude certain files and folders using rules specified in the ".gitignore" file, but by default choosing which files and folders to **track** is done manually. It is recommended to include a "read me" file such as **readme.md** (although this is not required). The **config** command also allows the customization of various settings, such as your name and email address and your preferred text editor.

Underneath the hood of all **repositories** is the **tree**. The **tree** is essentially the version control data structure. This accommodates non-trivial projects, where the development **tree** has different **branches**, allowing deviations, or multiple versions of the code to be developed concurrently. Each node on the **tree** represents a **snapshot**, or a saved version, of your code[2]. A **snapshot** is automatically generated when you create a new **repository** using the **init** command. To create a new **snapshot**, besides the very first, you first need to

---

[1]https://git-scm.com/book/en/v2

[2]Why is it called a "tree" you may wonder? This is because the tree is in fact a "K-tree" from the perspective of data structure theory. Also relevant is that trees have branches, and branching is a central feature when talking about version control systems.

stage the changes to be included in the **snapshot**, and then **commit** these changes. It is mandatory to include a **commit** message describing changes in git (otherwise good practice anyhow). How you should write descriptive and concise **commit** messages is beyond the scope of this chapter.

The **init** (initialize) command naturally also generates a **branch** for the first **snapshot** to reside on. It is called the **master branch**. The **master branch** sounds very intimidating, but in fact it is not special in any way beyond being the default branch created when a **repository** is first created. You can even change its name (but very few bother). It is entirely possible to utilize this **branch** only, but to do so would be to ignore a powerful feature of version control systems in general. **Branching** allows the creation of parallel versions of your code to co-exist in development concurrently. For example, a new **branch** can be created for development of new features or bug fixes and later merged into the **master branch**. Excessive **branching** may be confusing though, so a good tip would be to always check the **status** before you **commit**, to minimize the risk of **committing** to the wrong **branch**. Also keep in mind that **merging branches** that have undergone extensive development since they diverged, even with the help of the **diff** command, is usually a non-trivial task.

The full might of Git is revealed when deployed in a server/client architecture (commonly through services like GitHub, but you can also host your own Git server). A **remote** Git **repository** is hosted on a server, be it on your own server in your basement, or on another computer elsewhere (read: someone else's computer). Whenever you want to work on a **remote repository**, you may **clone** the **tree** (make a local copy). If the **remote repository** has already been **cloned** by you previously, you should **pull** the **remote tree** (or relevant **branch**) before working on it to ensure you have the latest version. Just do not forget to **push** what you have worked on to the **remote** when you are done, so others can start working from where you left off. A **master branch** of a **repository** always refers to its main branch on your local machine. To avoid confusion, the corresponding **branch** to your **master branch** on the **remote repository** is called **origin**.

This covers the essential concepts of Git (shared with most other version control systems, although the precise terminology may vary), as well as introducing some of the commands often used when using Git. With the jargon out of the way, we will say something about how to use version-control systems in MATLAB.

Version Control Systems: SVN and Git integration in MATLAB

MATLAB integrates with Git and Subversion (SVN). This integration provides the users an easy and convenient way to work with the source control within MATLAB. The full range of operations can be performed for MATLAB's Current Folder browser, such as, checking out from a repository, creating new repository, merging and resolving conflicts, committing and reverting changes, moving, renaming and deleting files.

All of the above mentioned version control operations are carried out as interactions in the MATLAB Desktop Environment. It is suggested to review the MATLAB documentation on setting up Git and SVN as well as its example workflows. Below is the MATLAB syntax for accessing this documentation:

```
doc git
```

and

```
doc svn
```

Note that SVN is built into MATLAB whereas you will have to install Git separately before starting to use it.

7.2.4 Testing

Software testing is a process of evaluating if a software program works as intended, a necessary condition of a good program. This topic comes in many forms, and is a science in itself. There are many dedicated books covering the many testing strategies, which is fortunate, as a more than cursory introduction is beyond the scope of this book. A common distinction made between testing strategies is *functional* versus *structural*. Functional testing evaluates the behaviour of your program and does not rely on knowledge of the source code (it is "black-box" testing). Testing whether an expected output is successfully produced by your program is functional testing. In contrast, structural testing does rely on knowledge of the code (it is "white-box" testing) and involves testing parts of it. Beyond simply saying here "Test your code!", we give a brief introduction to a simple testing method implemented in many established programming languages, including MATLAB: *unit testing*. Unit testing is considered to be a form of structural testing.

Unit testing

The functionality of software can be tested at multiple levels, for example: unit testing, integration level testing, system level testing, etc. Unit testing is the first level of testing, which as its name suggests, ensures that each individual unit of the code gives the desired outcome. To better understand unit testing, let us say you are developing a program for a basic calculator. The calculator should be able to perform addition, subtraction, multiplication and division. Functionality-wise, any operation can be combined with another. So, in this calculator example, unit testing implies testing each operation at individual level, for example, making sure that the function to perform addition gives the expected output given different inputs.

Manual vs. automated testing

Manual testing means testing a piece of software "by hand". In the above example of testing a function to perform addition, manual testing would imply, manually trying out different inputs to the addition function to check for the expected outputs. Although, in this example, manual testing sounds easy and straight forward, the task gets cumbersome when one needs to repeat the manual testing exercise every time one makes changes to the code. Besides, manual testing is prone to human error. This pain of testing manually every time and the vulnerability to human error can be alleviated by automated testing. In automated testing, we can write programs to test a piece of software. So, whenever a change is made in the software, by running the test program, any bugs can be discovered. Whether you should go for manual or automated tests depends on the return on investment. For that reason it is up to the programmer's discretion to choose one over the other.

Unit testing using MATLAB's testing framework

MATLAB provides a comprehensive testing framework for writing unit tests. Using the MATLAB Testing Framework, one can write script-based, function-based and class-based unit tests. This section explores a simple approach using script-based unit tests. To understand writing script based unit tests, let us consider the addition operation from the calculator example we considered earlier. Our goal is to write a script-based unit test to ensure that the function for addition returns the expected output.

Create `addition.m` function

We will begin with writing a function called `addition`. The type of inputs to the `addition` function is limited to a real scalar double value. Passing any other type of input should throw an error.

```
function result = addition(a,b)
% Check input types of a and b. Throw error if input is invalid
if(isa(a,'double') && isreal(a) && isscalar(a) && ...
        isa(b,'double') && isreal(b) && isscalar(b))
    result = a + b;
else
    error('One or both inputs are invalid. Input should be a real scalar double.');
end
end
```

Create test script to test addition.m

We will also create a test file `testAddition.m` in the folder where the `addition.m` file lives. We will test following test cases:

- **Test 1**. Both inputs are integers.

- **Test 2**. First input is a double and second input is an integer.

- **Test 3**. First input is an integer and second input is a float.

- **Test 4**. First input is a negative integer and second input is a positive integer.

- **Test 5**. Both inputs are negative integers.

It is important to remember here that MATLAB stores integer inputs as type double, unless the user explicitly uses one of the eight integer class types (e.g., `int32`). We will not use these integer classes, that is, the function `addition` should not raise an error for any of our five test cases. When writing the test file, we must ensure that the file adheres to following constraints:

- The test file name must have the word "test" at the start or at the end.

- Each unit test in the test file must be put in its separate sections. A section begins with two percent signs (%%). The test framework interprets the text following the (%%) signs as the name of the test point.

- Any code before the first section is interpreted as the shared variable section. Shared variables are the variables shared across different unit tests in the test script. One can modify any shared variable in a test point, however, the variable gets resets to its default value at the beginning of any new test point with the test file. Note that any variable defined in a test point is not accessible to another test point unless it is defined in the shared variables.

- Define any preconditions necessary for the test in the shared variable section (this can be using the `assert` function). MATLAB does not run any of the tests on failure of the preconditions defined.

```
% Test addition
set1InputA = 2;
set1InputB = 4;

set2InputA = 0.20;
set2InputB = 10;

set3InputA = 2;
set3InputB = 10.12;

set4InputA = -10;
set4InputB = 20;

set5InputA = -10;
set5InputB = -20;

%% Test 1: sum when both inputs are integers
result = addition(set1InputA, set1InputB);
assert(result == 6);

%% Test 2: sum when input 'a' is float and 'b' is integer
result = addition(set2InputA, set2InputB);
assert(result == 10.20);

%% Test 3: sum when input 'a' is integer and 'b' is float
result = addition(set3InputA, set3InputB);
assert(result == 12.12);

%% Test 4: sum when input 'a' is negative and 'b' is positive integer
result = addition(set4InputA, set4InputB);
assert(result == 10);

%% Test 5: sum when both inputs are negative integers
result = addition(set5InputA, set5InputB);
assert(result == -30);
```

Running the test

You can run the tests in **testAddition.m** using MATLAB's **runtests** function. It executes the test cases in **testAddition.m** separately. If any test case fails, MATLAB continues to run the subsequent test case. The **runtests** function reports the status of the run, including any failed or incomplete test. Typing:

```
testResults = runtests('testAddition')
```

will produce the output:

```
testResults =

  1x5 TestResult array with properties:

    Name
    Passed
    Failed
    Incomplete
    Duration
    Details

Totals:
    5 Passed, 0 Failed, 0 Incomplete.
    0.082361 seconds testing time.
```

You can create a table of the `testResults` to conveniently extract details from the results above using the following command: `table(testResults)`.

Writing advanced tests
We reiterate that a strength of such unit testing is that it allows convenient repeat testing. A framework like unit testing provides the programmer with a means of automatically verifying the outputs every time a significant change is made to the code.

The MATLAB Testing Framework also provides capabilities to write advanced tests based on functions and classes. However, a detailed treatment of advanced tests lies beyond the scope of this book.

7.3 CONCLUSION

Writing code that works and writing good code are two different things. When designing a program, try to design a good program from the very beginning. Consider using version control. Comment your code. Consider how you are going to test your code, to minimize the chance of nasty surprises. If you do all these things, you are likely to minimize your misery as well as that of others.

MATLAB toolboxes used in this chapter:
None

Index of the in-built MATLAB functions used:

assert	isa	isscalar	table
error	isreal	runtests	

## CHAPTER 8

Sharing software

Yanlu Wang

Department of Clinical Sciences, Intervention and Technology, Karolinska Institutet, Stockholm, Sweden
Medical Radiation Physics and Nuclear Medicine, Karolinska University Hospital, Stockholm, Sweden

Piyush Khopkar

MathWorks Inc., Natick, MA, USA

CONTENTS

T HIS chapter should be seen as a relatively free-wheeling introduction to sharing software, before the serious business of what to consider if you are deploying your software in a clinical environment where it could affect patients (Chapter 9) and attempting to conform to industry-standard practices for software development (Chapter 10). For the target audience of this book, we will assume you are interested in both sharing your software with colleagues, friends or the public and maintaining/updating your code according to user feedback. Hence aspects of both crowd-sourcing and open-sourcing will be briefly covered in this chapter.

You may wish to only share your code with a small number of collaborators or colleagues. Such "little crowds" do not constitute an open-source project, or even true crowd-sourcing. Some of the same considerations and tools are relevant, however. Of course, you may not wish to share your code at all, but rather share an *application* to provide a service to users. We have that covered in this chapter too.

8.1 POTENTIAL OF CROWD-SOURCING

To understand the potential of crowd-sourcing, one first needs to understand what it is. The definition of crowd-sourcing varies depending on the circumstances. The definition most commonly employed in the scientific, academic and community development domains is:

Crowd-sourcing: The act of obtaining information, or input, into a project by enlisting the services of a large community of individuals.

This is the definition we will use in the chapter, with the qualification that we will not assume that the community is particularly *large*. One can imagine the huge benefits of crowd-sourcing after reading the definition alone. The terms crowd-sourcing and open-sourcing are sometimes confused and used interchangeably, but they are not the same thing. Open-source is when you "freely" publish your code for everyone to see, use and change as desired, although there may be some restrictions on use under the terms of the license (see the licensing section). One can crowd-source fully proprietary software through assimilating the public's ideas, and provide open-source code without ever crowd-sourcing, through a specific choice in license and leaving the project unmaintained after release. There are two general approaches to take when crowd-sourcing.

Obtain ideas, suggestions and feedback from end-users
This approach is commonly employed by companies, as the source-code may be confidential and not published. Companies may also publish parts of a code for crowd-sourcing or publish an application program interface (API), for which one can freely write code and programs to interface with the company's proprietary code. Companies with yet-to-release software can also pre-release it to a select crowd, or the general public, for "beta-testing", in return expecting bug reports and suggestions for the upcoming release. On a much smaller scale, you may simply share a prototype application you have compiled from your code with colleagues and other end-users, for the purpose of getting feedback.

Invite others to inspect/modify your code
Crowd-sourcing can also be applicable for individuals, or small groups of developers working on a community project, where the project code is shared and possibly open-source. These projects not only accept bug reports, ideas, or suggestions to improve the project, but also welcome contributions in the form of source-code. Many developers are happy to recruit new contributors and co-developers in their project. Using version control and other software management tools (see the previous chapter: *Good programming practices*), this model scales from the level of collaboration between two colleagues at a single institution to a community of thousands on an open-source project. Another obvious strength of crowd-sourcing is increased exposure and publicity. Crowd-sourcing also has some secondary benefits beyond the obvious. In the case of sharing source-code, the very thought of others reading your code encourages good programming practices. It will encourage well-written documentation, something that you probably would not even bother doing if you were not intending others to see the code in the first place (and regret later on). Even writing an API forces one to think about the code structure. Simply receiving input and ideas, crowd-sourcing has the potential to develop your skills in maintaining code.

Whether it is about prioritizing feature requests on a forum, or sorting lists of bugs to fix, it will not only develop you as programmer, but also benefit your management skills.

8.2 SHARE CODE USING MATLAB FILE EXCHANGE

MATLAB provides an easy and convenient way to share the MATLAB code with the community through *MATLAB File Exchange*[1]. MATLAB File Exchange is a free service which allows users to upload files up to 20 MB. Files can be either uploaded from a computer, or can also be shared from a *GitHub* repository or by sharing the website containing your code. In addition to submitting the files, MATLAB File Exchange allows users to:

- **Search and download** files such as, demos, examples, MATLAB apps, *Simulink* models submitted by other users within the MATLAB community. Files can be searched using the file type, author name, author rating, tags (keywords), file rating, or license type. MATLAB treats file exchange submissions as "Add-Ons". Users can also search and download the MATLAB File Exchange submissions from the *Add-Ons* option of the MATLAB Toolstrip

- **Tagging the files**. Users can add tags or keywords to their own submissions, or to others' submissions as well

- **Rate and comment** the submissions

Below are the good practices to follow while submitting files to MATLAB File Exchange via the site:

- Ensure the code runs without errors
- Have a clear title and description
- Add a proper version number. Versioning helps to distinguish between updates made to the existing submission
- Add acknowledgements if your work is inspired by others' work and instructions for others to cite your work
- License your work appropriately. By default, all the submissions posted on the MATLAB File Exchange will be under the BSD license unless otherwise noted. Section 8.7 will cover licensing in detail
- Add relevant tags and keywords for better discoverability
- Add dependencies on other MathWorks products. For example, if your submission requires MATLAB's *Image Processing Toolbox*, spell that out in the "Required MathWorks Products" field
- Add appropriate information in "MATLAB Release and Platform compatibility"

By default MATLAB File Exchange zip all the submitted files. The submission can also be packaged as a *toolbox*. MATLAB will install the toolbox in R2014b or later versions (simply double-click on the downloaded .mltbx file in the *Current Folder* window of MATLAB).

8.3 SHARE CODE USING OTHER SOURCE-CODE HOSTING SITES

Beyond MATLAB File Exchange, there are plenty of independent source-code hosting sites, many of which are integrated with version control and can host your software as repositories. Well-known examples include *SourceForge*, *GitHub* and *Bitbucket*. Host sites offer differing charging models. Many provide free services (with limitations) for non-commercial users. There is also the option of hosting your own site, privately or at an institution, using software such as *GitLab*. We recommend the reader take a look at a few examples. Below is a list of repositories for software used elsewhere in this book:

---

[1]https://www.mathworks.com/matlabcentral/fileexchange/

- **Code repository for this book**. The MATLAB code accompanying this book[2]
- **Code respository for xrTk**. The MATLAB code for the xrTk software described in Chapter 16 (*A toolkit for x-ray physics calculations*)[3]
- **Code repository for SpekPy**. The Python code for the SpekPy software introduced in Chapter 6 (*Integration with other programming languages and environments*)[4]
- **Code repository for EtherJ**. The Java code for the EtherJ DICOM library discussed in Chapter 18 (*Image processing at scale by "containerizing" MATLAB*)[5]
- **Code repository for Evil DICOM**. The .NET code for the Evil DICOM package introduced in Chapter 6 (*Integration with other programming languages and environments*)[6]

Note that in most cases you do not *have* to make your repository public. You can keep it private and share it with select individuals.

8.4 CHOOSING THE OPTIMAL APPROACH: GUI OR NOT?

In what form should you share your software with other people? The optimal approach to take when designing software is primarily dependent on the task the program solves. This is true for all software, independent of platform or programming language. The term "approach" is intentionally vague in this formulation, but we are going to concern ourselves mainly with user-machine interface design decisions: specifically, whether or not to incorporate a Graphical User Interface (GUI). Consideration of the intended end-user is paramount and should be the starting point for all interface decisions. In some cases, this point is very straightforward, since the software is designed to solve a specific task in a specific way (for a specific individual or group).

Some programs naturally lend themselves to GUI. Examples of this are all types of visualization programs: image and video viewers, editors, data visualization programs such as CAD viewers, and publishing programs. One would be hard pressed to find such a program without at least some sort of GUI interface. Indeed, it is a challenge to just imagine a reasonable design for such programs without including at least one GUI component. On the other hand, some programs tend not to have a GUIs for equally good reasons. Some may think that the "lack" of a GUI is a sign of laziness from the side of the programmer. While this may indeed be the case at times, there are two things to take into account when judging a program solely on the basis of its interface. Firstly, every developer's time is limited, and incorporating a GUI is not a trivial task, even if one is already an experienced expert in implementing GUIs in a specific environment. The time taken to write GUI elements might easily be allocated to debugging existing code or writing additional functionality to the program and thereby extending its usefulness. Having a GUI does not necessary make a program better. Secondly, some programs are simply best run in the command-line. Some examples might be data crunching programs, application-specific programs intended to be kept in the background, and generally programs designed to a intermediary part of a larger processing pipeline. These programs are not meant to output anything to the user beyond warning and error messages, and usually expect a specific input format for its input data.

However, there is certainly a place for GUIs beyond the essential. The mere presence of a GUI, no matter the application, lowers the initial learning curve of your program

---

[2]https://bitbucket.org/DRPWM/code
[3]http://dqe.robarts.ca/icunningham/xrTk
[4]https://bitbucket.org/spekpy/spekpy_release
[5]https://bitbucket.org/icrimaginginformatics/etherj-core
[6]https://github.com/rexcardan/Evil-DICOM

dramatically. Those not comfortable in command-line programs might be dissuaded from using a program if all it offers is a list of command-line options. On the other hand, there is simply no way a GUI can perform certain tasks, at least not as elegantly and efficiently, as command-line calls. When a GUI is implemented for a program which is best suited to be called from a command-line, the GUI either offers access to only a fraction of the functionality, or the GUI is overwhelmingly difficult to operate and manage.

In the case of MATLAB programming, the decision whether to implement a GUI or not is not such a stark choice. MATLAB is easy to learn despite its origins in numerical computing and deep root in academia. Although most MATLAB programs can make do without a GUI for user input, many programs in MATLAB are intended to be a teaching tool where graphical GUIs are extremely useful. In addition, most MATLAB programs in the field of medical imaging make use of its extensive visualization capabilities (i.e., in-built GUI components such `plot` and `imshow`) to a certain extent. Many of the aforementioned points regarding user-interfaces still apply, however.

When considering the purpose of the program and the optimal approach to accomplish its purpose, the decision to include a GUI may come naturally. The intended audience of your program remains paramount. It makes little sense to force an experienced user to periodically have to click *Ok* during 8-week long processing jobs. On the other hand, educational programs for children should not assume the end-user is fluent at the syntax to use in the MATLAB *Command Window*.

8.5 BUILDING AN APP IN MATLAB

Once you have settled on the purpose and audience of your program and made the decision of creating a GUI, the next step is building the application. An *app* or an *application* is a self-contained MATLAB program that provides an easy and convenient access to the program. You can have interactive controls, menus, buttons and plots for data visualization in an app. In MATLAB, apps can be created using two workflows:

- **App Designer:** Introduced in R2016a, *App Designer* provides an interactive development environment with large set of components, integrated into the MATLAB editor. Additionally, applications created using App Designer can be easily packaged and compiled for sharing, either as a standalone desktop application or as a MATLAB web app

- **Programmatic workflow:** Through this workflow, an app is a traditional MATLAB figure with many interactive components placed in it and saved as a ".fig" file. The functionality of the app created from this workflow is equivalent to what can be created interactively within App Designer

The best choice among these will depend on your project requirement and your personal preference. Considering App Designer's simplicity, convenience and the interactive way of creating apps, in the rest of this chapter we will focus on GUI development using App Designer. In particular we will walk through the App Designer environment, the code and design view and its component library. Our hope is that the example app presented here will give you a head start in developing your own applications.

8.5.1 Open app designer

Invoke App Designer by running the `appdesigner` command in MATLAB command prompt:

```
appdesigner
```

You can also open an existing app by passing the filename as an input:

```
appdesigner <fileName>
```

8.5.2 App designer environment

Figure 8.1: The App Designer environment.

The environment in Figure 8.1 is split up into following main parts:

1. **Component Library**. This contains the interactive components. The *Components* subsection will discuss the *Component Library* further.

2. **Editing Space**:

    (a) **Design View** The design view is like a canvas for you to drag and drop the components you want in your program.

    (b) **Code View** Code view provides the access to the MATLAB code editor environment. You can switch between code and design view at any point while developing an app.

3. **Component Browser** As you drag and drop the components for your app, those components will appear in the component browser. Components in this browser will appear as a tree like structure showing their hierarchy. All the components of an app are placed on top of a base figure, `app.uifigure`, which will be the default base component in the component browser. Selecting any component from the Component Browser opens Property Inspector. One can inspect and modify component properties, such as color, title or labels through the Property Inspector.

8.5.3 Components

One of the salient features of App Designer is its large component library for designing modern and full-featured applications. The components are categorized into three main parts:

1. **Common Components**. This include commonly used components such as, `axes`, `text box`, `drop down box`, `table`, `checkbox`, `slider` and `date picker`.

2. **Figure Containers and Figure Tools**. These include the containers: `panel`, `tab group`, `grid layout manager` and figure tools: `menu bar`. The containers allow you to group the components based on the project's requirements. With the menu bar you can create menus such as: `File`, `Edit`, `View` or any other menu item as desired.

3. **Instrumentation Components**. Contains a large library for different instrumentation components, such as `gauge`, `knob`, `switch` and `lamp`.

8.5.4 Create and run a simple app

Now that we have familiarized ourselves with the App Designer environment and its component library, it is time to get our hands dirty. In this section we will create and run a simple app using one of the common components. Let us say the goal of this app is to show the number of CT scans in two clinics between 2015 and 2018, with the option of user-interaction regarding which data is presented. Basic idea: by default the app will show CT scan data from both the clinics. User can switch between clinics via the radio buttons on the right. We will break the app creation and the running of the app into a few steps.

Step 1: Create a rough sketch of the design
The ease of use of an app is completely dependent on how simple and intuitive the app is designed. Pencil and paper are good for scribing the design. Figure 8.2 illustrates a sketched design for our app.

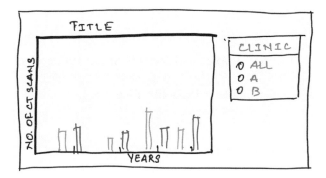

Figure 8.2: A rough sketch of the app.

Step 2: Lay out components in App Designer in Design view
Based on the app design you we scribed in step 1, we will now lay out the required components in the App Designer. We will open up App Designer and perform the following steps:

1. Drag an **Axes** and a **Radio Button Group** component from the component library's *Common components* section to the canvas

2. Select the **Axes** component by clicking on it (see Figure 8.3) and change the axes *Title*, *X Label* and *Y Label* properties to: *Number of CT Scans in Clinic A and B from 2015 to 2018*, *Year* and *Number of CT Scans (in hundreds)*, respectively

3. Select the **Radio Button Group** by clicking on it and change the text *Title* to *Clinic*, change the text of the buttons to *A* and *B* and delete the extra button.

Figure 8.3: **Axes** and **Button Group** Properties.

You should now have something that looks like Figure 8.4.

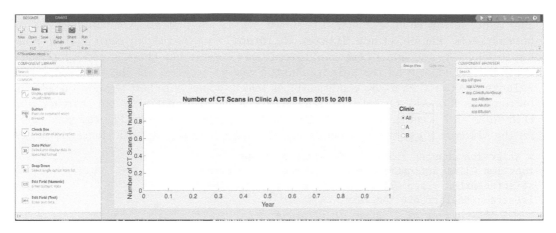

Figure 8.4: The *Design View*.

Step 3: Add callbacks

A *callback* is a MATLAB function that executes when a certain action is performed, be it on starting the app or selecting a component. For this app we will add two callbacks. First, a *startupFcn* callback that will get executed on running the app. Second, a *radio button* callback, that will get executed on selecting any of the radio buttons. Perform the following step for adding a *startupFcn* callback:

1. Switch to the code view if you are not there already

2. From the Code Browser select *Callbacks* tab and click on the + button. In the *Add Callback Function* window, select *UIFigure* in the Component dropdown and *StartupFcn* in the Callback dropdown if that is not selected by default. Finally clicking on *Add Callback* will create a *StartupFcn* callback

For adding the *radio button* callback, select and right click on the radio button and select *Callbacks* from the context menu. Select *Add SelectionChangedFcn callback*. Following this, App Designer will create an empty callback function and put the cursor in the body of the callback function. This will also switch the view from *Design* to *Code* view.

Step 4: Write callback code

Earlier we made the decision of showing the number of CT scans at both clinics on opening the app. For that we need to write the code for this part in the *startupFcn* callback.

```
function startupFcn(app)
% Create dummy data
y = [2 2; 2 5; 2 8; 11 12];

% Visualize data through bar chart
% app.UIAxes is an handle to the axes
bar(app.UIAxes,y)

% Set XTickLabels
app.UIAxes.XTickLabel=  ['2015';'2016';'2017';'2018'];
end
```

In above code snippet, **y** is the number of CT scans (in hundreds) at clinic A and B during 2015 to 2018. The code **bar(app.UIAxes,y)** plots two bar charts representing clinic A and clinic B.

The radio button callback function is shown below, but first an explanation of the different objects. The object **app.UIAaxes.Children(1)** represents bar chart 1, corresponding to clinic A, and **app.UIAxes.Children(2)** represents bar chart 2, corresponding to clinic B. The identity of the **selectedButton** can be accessed via the **Text** property of **selectedButton**. The *if-else* condition block selects which bar chart to show on selecting button **A**, or **B** or **All**. On selecting radio button **A** the code makes the bar chart for clinic B invisible and selecting **B** makes the bar chart for clinic A invisible. The *else* condition is for the **All** case where the code makes the both the bar charts for both clinics visible.

```
% Selection changed function: ClinicButtonGroup
function ClinicButtonGroupSelectionChanged(app, event)
selectedButton = app.ClinicButtonGroup.SelectedObject;
if(strcmpi(selectedButton.Text, 'A'))
    % Show number of CT scans in clinic A
    app.UIAxes.Children(1).Visible = 'on';
    app.UIAxes.Children(2).Visible = 'off';
elseif(strcmpi(selectedButton.Text, 'B'))
    % Show number of CT scans in clinic B
    app.UIAxes.Children(1).Visible = 'off';
    app.UIAxes.Children(2).Visible = 'on';
else
    % Show number of CT scans in clinic A and B
    app.UIAxes.Children(1).Visible = 'on';
    app.UIAxes.Children(2).Visible = 'on';
end
end
```

Step 4: Run the app

Click *Run* to save and run the app. Once saved you can also run the app again in App Designer by typing its name without the extension (.mlapp) at the MATLAB command prompt. Before running the app make sure the file is in the current folder or on the MATLAB path. By default the app will show the number of CT scans for both the clinics, as illustrated in Figure 8.5. Click on A or B to toggle the view to specific clinics.

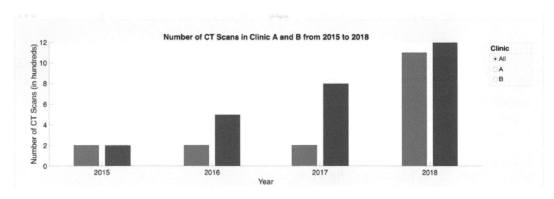

Figure 8.5: The *CTScanData* App.

8.6 CREATING EXECUTABLES WITH THE MATLAB COMPILER

The App Designer app we created in above section can be shared among colleagues, friends or clients by simply sharing the .mlapp file. Although sharing an .mlapp file is the quickest way to share the app, the .mlapp file exposes the source code to the end user and that entails a risk of modification of the app. Besides, the end user needs to have access to MATLAB in order to run the .mlapp file. An alternative workflow is to use *MATLAB Compiler*, which provides a convenient way of sharing apps as a standalone program.

The biggest advantage of a standalone program is anyone who does not have access to MATLAB on their system can run the program. The MATLAB compiler produces an executable that installs both the standalone application and all the dependencies required to run the program on the target machine. The target system does not need a MATLAB license for this purpose. Another advantage of a standalone program is that the source code remains hidden from the users of the application.

The App Designer provides an easy access to MATLAB Compiler for generating a standalone executable. You will need to have MATLAB Compiler installed on your system, so make sure that it is included in your license and has been installed. Follow below steps for creating a standalone executable of the app we created in the above section.

Step 1: Start the app. In 8.5.4 we created the *CTScanData* app. Open this app in the App Designer.

Step 2: Start MATLAB Compiler from the App Designer toolstrip. Click on *Share - Standalone Desktop App* (see Figure 8.6 below)

Figure 8.6: How to share an app.

Step 3: Set the compiler Options. Compile options can be found in the in the *Compiler* toolstrip:

- Main File: Ensure that the *Main File* section of the toolstrip has the correct main file for the application. In our example it is "CTScanData.mlapp".

- Packaging Options: This option lets you decide whether to include *MATLAB Runtime Installer* in the generated application package. The options include:

    1. Runtime downloaded from web: The *MATLAB Runtime Installer* will be down loaded from web at the time of installing the standalone application

    2. Runtime included in package: The *MATLAB Runtime Installer* will be included in the generated package

The former option downloads and installs the *Runtime* during installation and requires internet connectivity as well, whereas in latter option, the packaged file is bigger in size because the *Runtime* comes with it. So, it is a one time trade off between installation time and connectivity and the size of the packaged app. Which option to pick is totally up to the app author, but we go with option 2 here since the installation is then free from an internet constraint.

Step 4: Customize the package.

- Application Information: Furnish necessary information about the application, such as, author name, contact details, application summary and a brief description. You can also change the appearance of the application by selecting a custom splash screen icon for the application.

- Additional Installation Options: Edit the default installation path of the generated application on the target system.

- Files required for your application to run: Select any additional files required for the application to run. By default, the *Application Compiler App* will try to find and include the required files for the application.

- Files installed for your end user: The installer will include the following files on the end-user's target system:

  - Readme.txt
  - The executable for the standalone application
  - Splash screen icon image

- Additional runtime settings: Other platform specific options for the generated executable.

Step 5: Package. Click *Package* to generate the packaged standalone application. Check on the *Open output folder when process completes* option. Upon completion of the packaging, the generated output should contain the following directories:

- *for_redistribution*: Folder containing the files for installing the application and the MATLAB *Runtime*.

- *for_redistribution_files_only*: Folder containing the files required for redistribution of the application. This does not include the installer for *MATLAB Runtime*. For that reason, share these files to the end users who already have MATLAB or MATLAB Runtime installed on their target machines.

- *for_testing*: Folder containing the artifacts created by MATLAB Code Compiler. This includes all the files in *for_redistribution_files_only* folder plus files generated during the steps of packing the MATLAB files.

- *PackagingLog.html*: Log file generated by MATLAB Compiler.

Step 6: Install and Run the Generated Standalone Application. Install the generated application by executing the installer file and follow the instructions on the user interface. To run the installed application, locate the executable file ("CTScanData"' in this case) in your file browser and click as normal to run it (e.g., double-click). Alternatively, you can open a command-line terminal and navigate to the appropriate folder, before invoking the executable from the command-line.

8.7 LICENSES

Although the last of the section in this chapter, choosing a license should be your first concern when deciding to distribute your work. Giving out, or publishing, your software is essentially a legal decision more than anything else. Which license to choose is dependent on why you choose to publish your code, whether you wish to profit from your work in the future, and your personal stance on open-source software in general. This section briefly explains why this is important, how it need not necessarily take lots of effort, and which licenses to choose. For the sake of brevity, this section ignores some subtleties in the details. But for the beginner, and those curious in these matters, this section is a quick introduction. Before we begin, we stress that this does not constitute formal legal advice. If in doubt always consult a legal expert. Further, the reader should note that before you can license software, you need to be sure that it is your sole intellectual property. In some cases, even work carried out outside of your work hours and not directly related to your job description remains the property of your employer. Consult your employment contract.

First of all, why even bother with licenses? Why not just upload your code to a website or code-bank and leave everyone a "you're welcome" note? This is not recommended, primary due to reasons concerning liability and warranty. The act of publishing your program to the public-domain inherently comes with some responsibilities with legal implications. If your code can be traced back to you, and your code causes damage, whether it be economic, physical, or otherwise, you may be legally liable. Most licenses also include limitations of warranty and warranty disclaimers, ensuring that you are not liable for providing warranty to your published software if you do not desire to do so. Lastly, contrary to popular belief, software that is published without a license is *not* in the public domain and is in fact fully copyright protected (in most countries) unless an explicit waiver is included. Therefore, the software cannot legally be duplicated or distributed before the copyright term has expired and it falls into the public domain. You can write your own license, but we would not recommend this either, unless you know a lawyer, or are willing to pay a lawyer for such tasks. You can simply use an existing license written by people who probably thought more about this than you have, and very much like the concept of open-source software. They were kind enough to publish the licenses themselves as "open-source", so that you can freely use them at your discretion. Open-source licenses are those that comply with the open-source initiative[7], all of which allow the software to be freely used, modified and shared. The open-source licenses are approved though the Open Source Initiative's license review process. Most of the software you can download and freely use today falls under one of these licenses. Even those who are not typically concerned with such matters may have come across the names of some of these licenses often enough to have them engraved in your memory. These include: *Apache license (2.0)*, variants of the *BSD license*, *GNU General Public License (GPL)* and *MIT license*.

A typical misunderstanding is that you cannot sell open-source software. This is not true as all open-source licenses allow commercial distribution. In fact, the Open-Source Definition guarantees this (though practically it may be unwise to pick most of these licenses for commercial purposes). Using a common analogy, open-source software is "free" as in *free speech*, but not necessarily "free" as in *free beer*. Some of these licenses may prevent you from distributing the software in a way you intend to however, as some licenses do not allow proprietary (closed-source) software to be distributed that uses the open-source software. A good term to know is "copyleft" licenses, which require that any derivative code that is published must be published under the same conditions.

If you wish to keep things simple beyond all else, the MIT license is extremely concise. It

---

[7]https://www.opensource.org/

also lets people do almost anything they want with your code, including publishing closed-source versions. If you wish to keep all derivatives of your work open-source, in the spirit of sharing, the GNU GPLv3 would be your choice. Most open-source libraries and software are licensed through GNU GPLv3. While there are many other licenses with slightly different permissions and conditions, these two are your go-to licenses. You may also choose what is known as "The Unlicense", in which there is absolutely no restrictions imposed on the user. However, The Unlicense is still legally a license, meaning you are limiting your liability and warranty.

Once you have chosen a license, you will need to apply it to the software. Standard license text files can be found online such as *Choosealicense.com*[8]. How to appropriately apply the license to your project depends on the nature of the project and which license you have chosen. For example, projects with GUIs and html-based documentation should have links to a license text file in the interface or page. Generally, the more restrictive the license, the more elaborate are the requirements on license prompt and availability of the full license text. Below we give a brief example for a simple project sharing source-code via a webpage.

First, state the license type clearly on the project's front page (e.g., in a "README.md" file). There is no need to copy the entire license file here; just state the name and refer to the license text. This alone is insufficient for legal purposes though, as the software itself should include the license as well. Put the full license text in a file called "LICENSE(.txt)" included with the source code, and then include a short notice as a comment at the top of *each* source file. The comment *must* include copyright date(s), name of holder and name of license. The comment should also state clearly where to find the full text of the license.

Some licenses also include instructions on how to properly apply the license to your project. For example the GNU GPLv3 suggests to include the following at the top of each source file:

> *Copyright (C) <year> <name of author>*
>
> *This program is free software: you can redistribute it and/or modify it under the terms of the GNU General Public License as published by the Free Software Foundation, either version 3 of the License, or (at your option) any later version.*
>
> *This program is distributed in the hope that it will be useful, but WITHOUT ANY WARRANTY; without even the implied warranty of MERCHANTABILITY or FITNESS FOR A PARTICULAR PURPOSE. See the GNU General Public License for more details.*
>
> *You should have received a copy of the GNU General Public License along with this program. If not, see <http://www.gnu.org/licenses/>*

Notice that the text does not specify exactly the name of the file containing the license text proper, though it is usually named either "LICENSE" or something equivalently obvious. You can choose to specify this, but it is not required. In general, the comment notice need not look exactly like the example above, as long as it starts with the same notice of copyright holder and code update dates, states the name of the license, and makes clear where to view the full license terms.

---

[8]https://www.choosealicense.com/

8.8   CONCLUSION

Building apps in MATLAB and sharing them via the MATLAB File Exchange digital distribution service is one way to share software. Sharing compiled executables is another. There are further alternatives, such as maintaining an online repository, that allow more interaction and collaboration with your users. While a full overview of available distribution methods is outside the scope of this chapter, hopefully we have motivated you, and along the way provided some pointers regarding what to think about before clicking *upload*. Deciding to share your code is the first and most crucial step after all. So please do consider sharing your code. No matter how small the contribution, the world can surely only benefit from your work.

MATLAB toolboxes used in this chapter: *None*   Index of the in-built MATLAB functions used: appdesigner   bar   strcmpi

Regulatory considerations when deploying your software in a clinical environment

Philip S. Cosgriff

Former Head of Nuclear Medicine (retired), United Lincolnshire Hospitals, Lincolnshire, UK

Johan Åtting

Group Chief Information Security Officer and Group Data Protection Officer, Sectra AB, Sweden

CONTENTS

P ATIENTS expect a safe healthcare system. Manufacturers of medical products employ strict quality assurance processes and both the process and the products are required to fulfill national and international safety regulations. These regulations often also apply to in-house developed medical devices, such as in-house developed software. Further, information security regulations such as the GDPR in the EU and HIPAA in the US concern health information privacy. This chapter provides an introduction and overview of the current regulatory frameworks with a focus on the EU- and FDA-regulations, and how they may apply to in-house developed medical devices and to management of health information privacy. The section on Medical Device Regulations is authored by Philip S. Cosgriff and that on Health Information Privacy by Johan Åtting.

Important notice: The text in this chapter should not be taken as formal legal advice. It should be seen as guidance only. If in doubt seek qualified legal advice.

9.1 MEDICAL DEVICE REGULATIONS

9.1.1 Introduction to medical device regulation

The main objective of this book is to show how practical everyday problems encountered in a radiology department can be solved by staff writing their own software using the MATLAB programming language. This so-called in-house development approach is attractive from several points of view, but there are also risks and potential pitfalls. The purpose of this chapter is to focus on these risks, and how to minimize them.

It is known, largely anecdotally, that the approach to in-house software development in scientific health care departments (e.g., radiology, radiotherapy, medical physics, clinical engineering) is highly variable in terms of quality assurance (QA). Very few national surveys have been published and the few that have have shown disappointing results. For example, a Canadian survey of medical physicists found that less than 20% of departments reported any formal policy or written guidelines for QA, with most (70%) admitting to only an informal "general understanding within the department about how things should be done" [20]. A major concern to user-developers is failure to comply with relevant regulations, as this could have serious consequences for the developers themselves as well as the hospital in which they work. This chapter will concentrate on medical device regulations, but it should be noted that other laws relating to product liability and/or general health and safety may apply in situations where the medical device regulations do not [21].

The systems of medical device regulation in the European Union (EU) and the United States (US) are similar but there are important differences, particularly regarding device classification and compliance. Some general principles apply to both systems, but it is necessary to discuss each in detail in order to highlight the differences. From a regulatory perspective, there are essentially two issues to consider when developing software in a clinical environment. The first is to establish whether the type of software being developed is covered by medical device regulations and, if so, how to comply with the regulatory requirements.

9.1.2 Regulation in the European Union

In order to determine whether a particular software product is subject to EU regulation, we need to briefly consider the current position of European medical device regulation, and to then focus on the specific issue of in-house developed medical software. Medical device regulation within the European Union (EU) has recently changed, with the introduction of the Medical Device Regulations ("MDR17"), which came into force on 25 May

2017. These new regulations (MDR 2017/745) replaced both the general Medical Device Directive (93/42/EEC) and the related Directive on active implantable medical devices (90/385/EEC)[1]. There is a new separate regulation for in-vitro diagnostic medical devices (2017/746), which replaces the previous Directive (98/79/EC). The main changes and challenges faced by medical device manufacturers (under MDR17) has been summarized in a recent report[22].

The main point about the new EU medical devices regulation is that it is in fact a regulation, not a Directive, meaning that it became law in all member states as soon as it was passed by the European Parliament. There is a transition period of three years (i.e., until May 2020), during which time manufacturers of currently approved medical devices will be required to meet the requirements of the new regulations. Annex I of the regulations lists the general safety and performance requirements (GSPRs); a new term which replaces the "essential requirements" used in the medical devices directives (MDD). As will become evident, the whole regulatory issue of software used in a medical context hinges on the definition of a medical device, which is defined as follows in the new EU regulations[23]:

"medical device" means any instrument, apparatus, appliance, software, implant, reagent, material or other article intended by the manufacturer to be used, alone or in combination, for human beings for one or more of the following specific medical purposes:

- *diagnosis, prevention, monitoring, prediction, prognosis, treatment or alleviation of disease*

- *diagnosis, monitoring, treatment, alleviation of or compensation for, an injury or disability*

- *investigation, replacement or modification of the anatomy or of a physiological or pathological process or state*

- *providing information by means of in vitro examination of specimens derived from the human body, including organ, blood and tissue donations*

which does not achieve its principal intended action by pharmacological, immunological or metabolic means, in or on the human body, but which may be assisted in its function by such means.

It should be noted that the above definition was extended to include prediction and prognosis of disease, compared to that previously published in Directive 207/47/EC. If the device under consideration qualifies as a medical device it will then be classified according to the following system:

Class I	low risk
Class IIa	low-to-medium risk
Class IIb	medium-to-high risk
Class III	high risk

There are a number of detailed *classification rules* in the regulations to aid in the assignment of the correct category. Depending on the application and context, clinical software used in a radiology department will usually be placed in class I, class IIa or class IIb.

There are essentially two types of medical software that fall within the remit of MDR 2017/745. The first, simply referred to as *medical device software*, is software that is integral

---

[1]Directives 93/42/EEC and 90/385/EEC were both amended by Directive 2007/47/EC, which clarified the position of "standalone" software, defined software as an "active" medical device, and made software validation mandatory

to a medical device (e.g., embedded software controlling a drug infusion pump). The second type is software running on a separate computer that interfaces with the "parent" medical device and processes data received from it. This second type of "standalone" software is referred to as *software as a medical device* (SaMD) and has been formally defined (by the International Medical Device Regulators Forum, IMDRF) as "software intended to be used for one or more medical purposes that perform these purposes without being part of a hardware medical device". Furthermore, to be considered as SaMD, the software must "perform an action on data for the medical benefit of individual patients".

MDR17 compliance requirements for a medical device intended to be placed on the EU market are dependent on its classification, so it is important to ensure that the device is correctly classified. Compliance in the EU regulatory system is described in terms of conformity assessment procedures (CAP) and these are generally undertaken by Notified Bodies (NB), appointed by the Designating Authority (DA) in the relevant EU member state. This is in contrast to the system in the US where marketing approval of medical devices is centralized and is the sole responsibility of the FDA.

In-house developed medical software

In contrast to the EU Directive that it replaces, the new MDR17 includes a specific section (Chapter 2, Article 5) on *in-house manufacture and use* (IHMU), so in-house developed software is now clearly covered. Clause 4 states that "devices that are manufactured and used within health institutions shall be considered as having been put into service" (normally implying full regulatory compliance), but the next point states that "... with the exception of the relevant general safety and performance requirements set out in Annex I, the requirements of this Regulation shall *not* [emphasis added] apply to devices manufactured and used only within health institutions established in the Union, provided that all of the following conditions are met". These conditions (Article 5.5) are obviously crucial, so they are quoted here verbatim:

(a) the devices are not transferred[2] to another legal entity[3],

(b) manufacture and use of the devices occur under appropriate quality management systems,

(c) the health institution justifies in its documentation that the target patient group's specific needs cannot be met, or cannot be met at the appropriate level of performance by an equivalent device available on the market,

(d) the health institution provides information upon request on the use of such devices to its competent authority, which shall include a justification of their manufacturing, modification and use,

(e) the health institution draws up a declaration which it shall make publicly available, including:

- *the name and address of the manufacturing health institution;*

- *the details necessary to identify the devices;*

- *a declaration that the devices meet the general safety and performance requirements*

---

[2]"Transferred" means either provided free-of-charge or sold.

[3]"egal entity" is not defined in MDR17, so the generally accepted meaning is assumed. Namely, "an association, corporation, partnership, proprietorship, trust or individual that has legal standing in the eyes of law". A legal entity has legal capacity to enter into agreements or contracts, assume obligations, incur and pay debts, sue and be sued in its own right, and to be held responsible for its actions.

set out in Annex I to this Regulation and, where applicable, information on which requirements are not fully met with a reasoned justification therefor[e],

(f) the health institution draws up documentation that makes it possible to have an understanding of the manufacturing facility, the manufacturing process, the design and performance data of the devices, including the intended purpose, and that is sufficiently detailed to enable the competent authority to ascertain that the general safety and performance requirements set out in Annex I to this Regulation are met,

(g) the health institution takes all necessary measures to ensure that all devices are manufactured in accordance with the documentation referred to in point (f), and

(h) the health institution reviews experience gained from clinical use of the devices and takes all necessary corrective actions. Member States may require that such health institutions submit to the competent authority any further relevant information about such devices which have been manufactured and used on their territory.

Member States shall retain the right to restrict the manufacture and the use of any specific type of such devices and shall be permitted access to inspect the activities of the health institutions.

This paragraph shall not apply to devices that are manufactured on an industrial scale.[4]

The above conditions, plus the relevant Annex I requirements, have been collectively referred to as the "health institution exemption" (HIE) by the UK's Competent Authority (the MHRA) in its recently published draft guidance[25].

It has been suggested[26] that Article 5.5b may cause significant problems in some hospitals due to the phrase "manufacture *and* use" [emphasis added] in relation to the quality management system (QMS) requirement. As it stands, in addition to the QMS implemented by the staff that developed the software, the clinical department in which it is *used* must also operate under an "appropriate quality management system". Many clinical radiology departments now operate under a recognized QMS, but this something that needs to be considered prior to device release/clinical use. Also, if the finished software is handed over to staff in another department (within the same hospital/health institution) some post-deployment surveillance and clinical follow up must take place.

Article 5.5f is less prescriptive than the Technical Documentation requirements set out in Annex II of MDR17, but should be addressed by means of a *Technical File*. This is essentially a complete project file that should contain information on specification, design, device classification, risk assessment (e.g., using ISO 14971), a GSPR checklist, test protocol, a post-release surveillance plan and instructions for use. It should also include a justification of the decision to opt for an in-house development project in this case (ref: Article 5.5c); information that may also need to be supplied to the Competent Authority (CA) on request (ref: Article 5.5d).

Compliance with the so-called HIE requirements will require the in-house development team to be well versed in software engineering techniques and relevant regulations, and for the employing health institution to invest in the ongoing training of those staff involved. In summary, although the application of the 2017 EU regulations to IHMU has been referred to as "light touch"[27], the new demands placed on in-house developers are considerable.

Having established that in-house developed medical software is covered by the new EU medical devices regulations, it simply remains to be determined whether a particular

---

[4]The meaning of "industrial scale" is unclear and has been the subject of some legal debate[24].

software application meets the definition of a medical device and, if so, to determine its classification. Unlike commercially developed medical devices, the compliance requirements for SaMD intended for in-house use only are the same for all classes of device. However, professional standards and best practice considerations would indicate more extensive testing and more detailed documentation for higher class devices. If the developer chooses to implement IEC 62304 (as is recommended) the need for more extensive testing of higher risk devices will be required anyway, through the (separate) concept of software safety class. Although not mandatory under the HIE, the need for Notified Body involvement (in an advisory capacity) should also be considered for class IIb and (especially) class III devices[26].

In addition to supplying justification information to the national Competent Authority *on request*, the in-house manufacturer may be required to register and perhaps submit an annual return describing its activities. The Competent Authority has the *right* to inspect the in-house developer's department, but this is thought unlikely for Class I or Class II products intended for internal use only[27].

In summary, as far as in-house produced medical software is concerned, the main change brought about by the new EU medical devices regulations is that developers are now legally required to employ appropriate quality management and risk management systems, even if the software to be developed is intended for internal use only. As under the previous Medical Devices Directive, if the SaMD is transferred to another legal entity then the full CE marking process is required.

Use of international standards

Although standards are essentially voluntary in the context of MDR17, their adoption represents the simplest way of demonstrating compliance, since this (generally) confers a "presumption of conformity" with the relevant statutory requirements, meaning that products resulting from the development process are (automatically) deemed to be safe. This is the "good process, good product" philosophy, which is the implicit basic assumption of all quality management systems.

The EC recently updated its list of *harmonized* European standards in support of the original MDD[28], and most of these remain relevant to the new MDR. The vast majority relate to specialized medical equipment hardware, but a few relate to medical software production. These standards are listed below with full titles for reference:

Quality management:

- EN/ISO 13485:2016. Medical devices—quality management systems—requirements for regulatory purposes
- EN/ISO 14971:2012. Application of risk management to medical devices
- ISO 9001:2015. Quality management system—requirements
- IEC/TR 80002-1:2009. Medical device software—Part 1: Guidance on the application of ISO 14971 to medical device software
- IEC/TR 80002-2:2017. Medical device software—Part 2: Validation of software for medical device quality systems

Process standards:

- EN/IEC 62304:2006[5]. Medical device software—life cycle processes
- EN/IEC 62366:2007[6]. Medical devices—Application of usability engineering to medical devices

Product standards:

- EN/IEC 60601-1:2006/A1:2013[7]. Medical electrical equipment—Part 1: General requirements for basic safety and essential performance
- IEC 82304-1:2016. Health software—Part 1: General requirements for product safety

Other standards:

- IEC 12207:2017. Systems and software engineering—software life cycle processes
- ISO 90003:2018. Software engineering—Guidelines for the application of ISO 9001:2015 to computer software

Radiology departments should seek certification to adopted standards as a means of proving compliance to either the national CA or NB as appropriate. Further discussion on how the above standards complement each other in the context of in-house medical software development can be found in a recent publication in by the Institute of Physics and Engineering in Medicine (IPEM)[29].

In house medical software developers in the UK are advised to follow the advice given in this chapter, which is based on EU MDR17. Post Brexit, specific UK medical device regulations will be published in due course, but are likely to be closely aligned with MDR17.

9.1.3 Regulation in the United States

The regulation of medical devices in the US is broadly similar to that in the EU, but there are some important differences in the way that manufacturers gain approval to market their new devices. Devices are regulated under the Federal Food, Drug and Cosmetics Act (FD&C Act) as amended by the Medical Device Amendments 1976 and subsequent related amendments; the actual regulation of medical devices under this statute falling within the remit of US Food and Drug Administration (FDA). The FDA generally classifies medical devices based on the risks associated with the device, and by evaluating the amount of regulation that provides a reasonable assurance of the device's safety and effectiveness[30]. The US system has three classification levels (Class I, Class II, Class III), with Class I being the lowest risk and Class III the highest. Unlike the EU system, the risk "labels" (low, medium, high) are not formally attached to the classes, but the basic control measures are as follows:

---

[5] IEC 62304 was amended in 2015, the revised standard referred to as IEC 62304:2006/A1:2015.

[6] IEC 62366: 2007 was replaced by EC 62366-1:2015, accompanied by a Technical Report (IEC/TR 62366-2: 2016. *Guidance on the application of usability engineering to medical devices*).

[7] The requirements for programmable electronic medical systems (PEMS) are now covered in Clause 14, replacing the previous collateral standard IEC 60601-1-4.

Class I: General Controls

- With Exemptions
- Without Exemptions

Class II: General Controls and Special Controls

- With Exemptions
- Without Exemptions

Class III: General Controls and Premarket Approval

The compliance system in the US is similar to that in Europe, with the degree of oversight dependent on the classification of the device. However, unlike the EU system, the FDA itself is responsible for the approval of new medical devices, either through the so-called 510(k) programme[8] or the more stringent premarket approval (PMA) system. The 510(k) process is a system of pre-market *notification* (PMN) that requires the manufacturer to demonstrate (by a written application) that the device in question is "substantially equivalent" to a legal device already on the US market—what the FDA calls a predicate device.

Class I devices are generally exempt from premarket notification procedures, but the manufacturer must comply with applicable FDA regulations, which includes the adoption of various "general controls" such as good manufacturing practice (GMP) techniques.

Class II devices require the implementation of both general and some "special controls", as well as a 510(k) pre-market *notification* submission.

Class III devices represent the highest risk to the patient and as such generally require the manufacture to undertake the full PMA process. In addition to the usual *process* controls deemed adequate for Class II devices, the applicant must provide the FDA with sufficient *validated scientific evidence* to assure that the device is safe *and* effective for its intended use[31]. Manufacturers must demonstrate to the FDA the effectiveness of these controls in providing a reasonable assurance of both safety *and* efficacy. This latter point highlights an important and fundamental difference between the US and EU approval systems for higher risk devices. In contrast to the US system, the EU approvals process only requires a demonstration of safety.

Regulation of Class I ("lowest risk") medical devices by the FDA is currently under review following the publication (in December 2016) of the 21st Century Cures Act, in which the previous FD&C Act definition of a medical device was amended. The change of definition (essentially removing some types of software previously defined as medical devices) necessitated that the FDA update some of its guidance documents and this process is still ongoing at the time of writing. The changes mainly concern types of health-related software that will no longer be defined as medical devices and thus will no longer be regulated by the FDA. This will include most types of clinical decision support (CDS) software, "general wellness" apps (e.g., fit-bit type) and what the FDA calls *medical device data system* (MDDS) software—that is, software that "passively transfers, stores, converts or displays existing medical device data"—bringing the US system more in line with the EU Regulations.

The situation with CDS software is contentious because the FDA has not previously issued specific guidance on this subject[32]. The statutory definition of CDS software in the *Cures Act* is very wordy and all three parts must apply if the software is to escape regulatory control. However, in the unlikely event that CDS software was being *totally relied upon* to

---

[8]The "510(k) process" simply refers to the section of the FD&C Act that deals with pre-market notification.

make clinical management decisions about an individual patient then that software would be deemed intrinsically more risky and may therefore be deemed to be a medical device. The way in which the FDA proposes to deal with the wider question of how the CDS software vendor *intends* its software to be used (by a clinician) has caused concern in both legal[33] and industry[34] circles and is something that the FDA will need to address/clarify.

Despite anticipated changes to the way in which low risk medical software is regulated, no changes are expected for "analytical" radiology/PACS software (i.e., software used in the diagnosis or treatment of an individual patient), which will continue to be defined as a medical device. Details of radiological equipment already classified by the FDA can be found on its respective PMA or 510(k) databases and general guidance is also available on premarket approval for different radiological imaging modalities[35, 36] and PACS systems[37]. Most radiological image analysis software is classified as a class II medical device and the FDA is proposing re-classifying some higher risk software (e.g., mammography, assessment of lung nodules) from class III to class II[38].

In-house developed medical software

In-house development of medical software is not specifically addressed in US regulations but it is clearly understood that developers are subject to the same rules as a commercial manufacturer if they intend to place their product on the US market. However, if the software is developed purely for use within the developer's own institution, one of the general registration exemptions referred to in the relevant Code of Federal Regulations (CFR) may be applied. Namely, "licensed practitioners, including physicians, dentists and optometrists, who manufacture or otherwise alter devices solely for use in their practice"[39]. In this situation, there is no legal obligation for in-house developers to employ GMP/software validation techniques, but most do so for professional reasons[40]. The current situation in the US is therefore essentially the same as it was in several EU countries under local implementations of the outgoing EU medical device directives.

Compliance issues

The majority of software defined as SaMD will be classified as either a Class I or Class II medical device, so will not need to go through the full PMA procedure. Most applications for FDA approval will be via the less onerous 510(k) pre-market *notification* route, but identifying a suitable predicate device may not be straightforward in all cases. Fortunately, the "substantial equivalence" sought relates mainly to the *intended use* (overall purpose of the device) and *indications for use* (particular conditions, clinical applications and environment in which device is to be used) rather than a side-by-side comparison of the respective device specifications.

The predicate device must thus be "broadly similar" to the new device in type and clinical application, but by no means identical. However, if the new device contains a substantial amount of new technology a question will be raised as to whether its inclusion could significantly affect the device's safety or effectiveness[41]. In a SaMD context, this might be the use of a new programming language that has not been previously used in this context. If the FDA approves the application, a "510(k) clearance" is said to have been awarded.

Formally, permission is then given to market the device, subject to compliance with relevant general and special control provisions of the FD&C Act (listed in the approval letter), since these will not have been inspected or verified by the FDA. A 510(K) clearance does not therefore mean that the FDA has determined that the device in question complies with other aspects of the FD&C Act, since that responsibility is left with the manufacturer.

Nearly all existing diagnostic radiology equipment (CT, MR, US, etc.) has been qualified as Class II and manufacturers of similar new equipment can thus make use of the 510(k) notification process. As far as PACS software is concerned, systems that simply store or transfer medical image data (without use of irreversible compression) are "Class I exempt" (i.e., exempt from 510(k) notification) whereas software that processes (changes) the data to aid interpretation is deemed Class II. This includes PACS mobile apps used for primary diagnosis[42].

The FDA has formally recognized several international standards (e.g., IEC 13484:2012, IEC 62304:2006/A1:2015, ISO 14971:2012) as a means for demonstrating compliance with some of the relevant FDA premarket submission requirements. As with EU regulations, certification to a particular standard does not guarantee compliance with the relevant part of the regulations. For example, IEC 13485 covers most of the requirements of FDA 21 CFR Part 820 (quality systems regulations) but not all[43].

9.1.4   Transfer of medical devices between different jurisdictions

Transfer of medical devices between different jurisdictions (e.g., EU and the US) should become simpler in the future as common schemes[9] and mutual recognition schemes are developed, generally mediated by the International Medical Device Regulators Forum. For now, however, the equivalence is mainly at the supporting international standards level, by formal recognition of certain key standards (e.g., IEC 13485, IEC 62304) as a means of demonstrating compliance with national or EU legal requirements for medical devices.

In-house software developed at a hospital in the US could only be transferred to a hospital in the EU if it met all the MDR17 regulatory requirements. This is analogous to a commercial medical device manufacturer in the US setting up a system to export devices to the EU. The receiving EU hospital would need to satisfy itself that all the Article 5.5 and relevant Annex I requirements had been met, and be able to demonstrate this to their national Competent Authority. Alternatively, the recipient could, by suitable written agreement, assume full ownership of the software and put it through its own software quality assurance procedures (which would obviously require access to the source code) to ensure that it meets EU regulatory requirements. The receiving hospital would then become the manufacturer and assume full responsibility for its use under EU law. Note that use of quality/risk managements systems and software development process standards is currently voluntary for software developed purely for in-house use in the US.

If the US hospital had received 510(k) marketing clearance from the FDA it could of course transfer/sell its software to another US hospital or private enterprise, but not currently to the EU, nor most other international jurisdictions. As the (class II) software was developed to higher standards (than in a non-regulated environment), it may well meet most of the MDR17 HIE requirements for in house use, but this cannot be assumed. The most likely route in this case would be for the US hospital (obviously experienced in medical device regulatory matters) to go through the full CE marking process, which would give it complete access to the EU market. Conversely, an EU hospital department that had successfully achieved the CE mark for its software device could not place it on the market in the US unless it had also gone through an appropriate FDA approval process.

In-house developed software in the EU can currently be transferred to a US hospital (for its own use, under the control of a named local licensed practitioner) as the device will be exempt from FDA regulation. From May 2020 onwards, all in-house software developed in the EU will have to meet MDR17 regulations, so US recipients after that date would get greater assurance regarding the performance, reliability and safety of the software.

---

[9]See, for example, the Medical Device Single Audit Program (MDSAP)[44].

9.1.5   Use of commercial off-the-shelf software

Nearly all medical software is constructed within the framework of a "programming environment" containing a high level language, library modules, and run-time functions. Suitable development packages (e.g., MATLAB, Maple) can be obtained from commercial suppliers or, alternatively, from open source providers (e.g., Python, IQWorks, OpenCV). Elements of these systems are thereby integrated into user-written software, so the question arises as to whether these "ready-made building blocks" need to be QA-checked by the developer?

The general issue of quality assurance of off-the-shelf (OTS) software used in medical devices was first highlighted in an FDA guidance document[45], which made it clear that the medical device manufacturer assumes full responsibility for software assembled using third party components. It defined OTS software (OTSS) as "a generally available software component, used by a medical device manufacturer for which the manufacturer cannot claim complete software life cycle control".

IEC 62304:2006 later introduced the term "software of unknown provenance" (SOUP), describing it as *a software item that is already developed and generally available and that has not been developed for the purpose of being incorporated into the medical device (also known as "off-the-shelf software") or a software item previously developed for which adequate records of the development processes are not available*. All OTS software was thus effectively defined as SOUP, with a number of measures prescribed to ensure that such software is suitable for use in a safety related environment[46]. The FDA/IEC definitions of OTSS and SOUP are roughly equivalent, even though commercial OTS (COTS) software is not necessarily SOUP or vice versa[47].

If unmodified open source software is used in a software medical device it should be treated as SOUP, and if changes are made to the source code (however minor) they should then be treated as a software item that you developed yourself[47].

MATLAB library functions/modules/runtimes clearly represent SOUP *by definition*. MATLAB is a high level scripting language so the finished medical application will be *mostly* SOUP; the user-written elements used mainly to link the pre-existing library elements into a unique configuration.

On the positive side, compliance with MDR17/IEC 62304 may not be as burdensome as using some other programming environments since MATLAB represents an example of what is referred to as "clear SOUP", for which the COTS-supplier has made available the development methodology, as well as independent reports from standards auditing bodies[48].

MathWorks provides a lot more information on their development processes than most commercial software producers, probably because it appreciates that this information might be required by developers using their software in a safety-related environment. Quoting directly from the MathWorks web site on the topic of "FDA software validation"[49], *"MathWorks can provide 1-page quality statements that describe quality driven development processes for the core platform products, MATLAB and Simulink. There are also detailed audit reports from a third-party independent testing body, TÜV SÜD in Germany. These are in regards to tool certification requirements of IEC 61508 standard[10], and attest that the software development and validation practices followed by MathWorks adhere to the highest standards in the industry"*.

---

[10]IEC 62304 is effectively a medical sector derivative of IEC 61508 (Functional Safety of Electrical/Electronic /Programmable Electronic Safety-related Systems). IEC 61508 Edition 2 (2010) is generally regarded as a higher standard than IEC 62304 and is typically used in industries that a truly safety-critical (mass transport systems, nuclear power stations, etc.)

9.1.6 Best practices

It should be remembered that standards essentially comprise a set of minimum requirements, and adoption of a particular standard, or standards, does not necessarily constitute best practice. The employment of "best practice" techniques by in-house software developers is clearly desirable for purely professional reasons, but it is also referred to in Annex 7, section 4.10 of MDR17 in relation to quality management systems.

A generally accepted definition of *best practice* is "a method or technique that has consistently shown results superior to those achieved with other means, and that is used as a benchmark"[50]. In this context, best practice constitutes the adoption of relevant standards and the use of associated guidelines on regulatory compliance. The use of appropriate *harmonized* standards certainly makes compliance with MDR17 more straightforward, and may also limit legal liability[11] in the unlikely event of a product liability claim being made against the developer/employing hospital.

A full discussion of relevant guidelines is outside the scope of this chapter but the following list provides a brief summary of what is available from professional bodies/national authorities:

EU MDR17 compliance:

- European Commission guidance documents relating to the new Medical Devices Regulations[51]
- European Commission official guidance documents (MEDDEVs) relating to the outgoing Medical Devices Directives[51]
- UK MHRA guidelines relating to medical devices[52]
- Sweden's Medical Products Agency guidance on "qualification and classification of standalone software with a medical purpose"[53]. *(This guidance document relates to the Medical Devices Directive, but it contains many useful explanations of medical device software terminology)*

US medical device regulation compliance:

- FDA guidance documents relating to the 510(k) submission process[54]

Quality management systems:

- IMDRF guidelines on QMS for medical device developers[55]

Software validation:

- FDA guidelines on software validation[56]
- UK National Physical Laboratory (NPL) guidelines on software validation[57]

Coding:

- Generic best practice guide[58]
- MATLAB programming/style guides[59, 60]

It is clear that MDR17 represents a substantially increased administrative and management overhead for all hospital departments undertaking software development work, and especially for those undertaking a few relatively small projects per year. The risks/benefits of this activity should thus be reassessed at senior department level and a policy decision taken

---

[11]For more information, search for "Development risks defence".

on whether to continue–probably with increased staffing resources–or cease. This review should include reference to three implementation guides recently published on the EC's "regulatory framework" website, under the heading "Information for manufacturers"[51]. The guides are aimed mainly at established medical equipment manufacturers who already have CE marked devices on the market, but they give a good idea of how much work is involved in preparing for the MDR17 implementation deadline in May 2020. Although some aspects will not be required for in-house developed devices, it has been suggested[61] that teams initially proceed as *if* they were intending to achieve CE marking.

Employing best practice techniques is obviously beneficial from many points of view, but it does not render the manufacturer/employer immune from prosecution in the event of a patient being harmed as a result of a defect in its software. If the software is *controlling* a radiation-producing imaging device (e.g., CT scanner) then the potential for harm (radiation over/under-dose) is obvious, but in-house developed radiology software is usually concerned with off-line image processing. In this case, the potential for harm is *indirect* and would involve a fault in the software leading to an incorrect diagnosis (false positive or false negative) that was then acted upon by the referring clinician. In this unlikely scenario, several competent practitioners would have to fail to detect the software error for the erroneous result to (a) be incorporated into the radiologist's report and (b) lead to inappropriate treatment of the patient. In the US, the FDA has long used the degree of "competent human intervention" (CHI) to help determine the classification of medical devices[62] and employing "control measures external to the software" (e.g., a "health care procedure") is permitted within IEC 62304:2006/A1:2015 as a means of reducing the software safety classification of the device, and thus potentially reducing the amount and rigour of software QA needing to be applied.

9.1.7 Radiology examples

According to MDR17 classification Rule 10, "active devices intended for diagnosis or monitoring" are classified as class IIa, but devices used in diagnostic or therapeutic radiology that *emit ionizing radiation* (e.g., CT scanners and diagnostic X-ray sets) are classified as class IIb. Radiological equipment that does *not* emit ionizing radiation (US, MR, NM) is therefore classed as class IIa.

Standalone software used in radiology that qualifies as SaMD would typically be classified as class I or class IIa, but (under *Implementing Rule* 3.3) software that "drives or influences the use of a medical device" automatically falls into the same class as the parent device. In radiology terms, this would usually mean software that controls or influences patient data acquisition by a scanning device.

Under Classification Rule 11, there is potential for diagnostic medical software to be classified as class IIb (or even class III) *in its own right* if used in critical situations where an inappropriate clinical management decision (heavily influenced by a diagnostic report) could lead to the death of a patient, or a "serious deterioration" in their health. Although this is theoretically possible with most types of image processing software, the link between software output and treatment decision in most clinical situations is (a) very indirect and (b) subject to numerous checks and comparisons with other tests. As a result, most diagnostic image processing software would be class IIa.

PACS software
PACS software receives special attention in the most recent official guidance on "borderline" medical devices, produced by a working group chaired by the European Commission[63]. It is important to stress that this "manual" (published in December 2017) refers to the

medical device *directives*, but in areas where the directives and regulations remain aligned (e.g., PACS software) the guidance is still applicable.

If the PACS software is intended only for data archiving and storage (using a method that preserves the original data) it would not be considered a medical device. If the software is intended for viewing, archiving *and* transmitting of patient data it is considered a Class I medical device. At the next level up, software intended to improve diagnostic accuracy by some form of *post*-processing (variously described in the regulations/guidelines as "altering the representation of data for a medical purpose" or "allowing direct diagnosis"[12]) is classified (under Rule 10) as Class IIa, partly due to the fact that all software is defined to be an "active device"[51]. Examples would include image filtering, multi-planar reconstruction and all quantitative assessment.

Radiation protection and equipment QA software

As far as radiation protection is concerned, software used to simply monitor a patient's cumulative radiation dose (e.g., from numerous CT scans) would not be considered a medical device *unless* the output of the software was used to influence the absorbed dose resulting from future procedures, in which case an error in the software could have an impact on the patient's health and would therefore be considered as SaMD. An example would be in-house produced "dose tracking" software originally developed for research use (so not SaMD by definition) that was subsequently used to compare local dose indices (e.g., dose-length-product in CT) with other hospitals, which might lead directly to changes in protocols or procedures at the originating center. If this new application was intended (i.e., documented in a revision of the software's *instructions for use*, IFU), it would then need to be re-classified as SaMD, and its whole development path reviewed. If it was being used inappropriately (i.e., application not described in the IFU) then responsibility for this would rest totally with the department that misused it.

The same basic principle applies to software designed to calculate the absorbed dose to individual organs from CT scanning or other procedures. If the program was used to halt or modify future examinations when the cumulative dose reached a certain level (e.g., for possible deterministic effects) then the software would be considered SaMD, as would be the case if the software was used to help decide which patients to refer to a dermatologist with a view to treatment for a suspected over-dose. Note that both of these qualifying provisions assume that such applications of the software were clearly stated in the IFU documentation. Software used to measure/monitor radiation doses to staff is not covered under these medical device regulations, but may be covered under ionizing radiation regulations.

Software designed for equipment QA purposes would not generally be considered SaMD, as the results are not applied to individual patients. For example, if a scanner was temporarily taken out of service due to the appearance of a computer-enhanced QC image (e.g., a poor uniformity parameter) then the clinical service might be disrupted (causing delays to a number of clinical procedures), but software used to prompt the action would not fulfil the criteria of a medical device.

However, if the same code were used for optimization of an acquisition protocol (e.g., calculation of the change in mA required to obtain a target noise level in the image) then it may be considered SaMD since the quality of the patient data obtained clearly has a direct effect on diagnostic accuracy, and a change in an acquisition protocol would affect subsequent patients. The key issue is the directness of the link between the software output and the decision to change the scan protocol. If the decision is ultimately made using

---

[12]A device is considered to allow "direct diagnosis" when it provides the diagnosis of the disease or condition by itself or when it provides decisive information for the diagnosis (MDR17, Annex VII, Chapter 2, 3.7).

some other method (e.g., by experimental verification of the target noise level using a physical phantom) then the aforementioned "QA software" would not qualify as SaMD. The crucial importance of the intended use of the software (which must be clearly stated in accompanying documentation) is hereby emphasized.

9.1.8 Conclusion regarding medical device regulation

The role of the in-house software developer has changed dramatically over the last 20 years, as it has become more of a recognized core duty than an interesting side-line for medical, scientific and technical staff. In the past it was also often performed by a single member of staff who just happened to be a skilled computer programmer, usually self-taught with no formal qualifications in software development. Increased awareness of the potential safety implications of faulty medical software has been accompanied by peer pressure to shift the paradigm from computer programming to software engineering, but this requires far more time and resources.

Until recently, software developed "for departmental use only" was exempt from medical device regulation (in the US and much of the EU) but the situation in Europe recently changed with the publication of the 2017 Medical Device Regulations. Such medical devices are spared the full rigour of the CE-marking process, but departments must now comply with specific regulatory requirements, including the implementation of a software quality management system and other best practice measures, which will need to be in place before May 2020. The minimum requirements for in-house medical software development under MDR17 are:

Work environment conditions and standards:

- Departmental policy on in-house medical software development
- Departmental coding and style standards for programming language(s) employed
- Public declaration of development activity (e.g., on the health institutions' website)
- Certification to IEC 13485 (Medical devices—Quality Management Systems—requirements for regulatory purposes)
- Detailed knowledge of ISO 14971 (Application of risk management to Medical devices)
- Certification to IEC 62304 (Medical device software—life cycle processes)

Individual software project deliverables:

- Technical file
- Administrator's file—describing how the code itself is stored, edited and protected
- General Safety and Performance Requirements (MDR17 Annex I) checklist—showing how each of the relevant clauses is covered

Sharing or externally marketing in-house developed medical software brings with it many more risks and responsibilities and it is thought that only the largest and better resourced radiology departments will be in a position to consider this option.

List of abbreviations for the first section of this chapter:

CAP	Conformity Assessment Procedures
CDS	Clinical Decision Support
CE	Conformité Européenne
CFR	Code of Federal Regulations
CHI	Competent Human Intervention
COTS	Commercial Off The Shelf
DA	Designating Authority
EC	European Commission
EN	European Standards
FDA	U.S. Food and Drug Administration
FD&C	Food, Drug and Cosmetics Act
GMP	Good Manufacturing Practice
GSPR	General Safety and Performance Requirements
HIE	Health Institution Exemption
IEC	International Electrotechnical Commission
IFU	Instructions For Use
IHMU	In-House Manufacture and Use
IMDRF	International Medical Device Regulators Forum
IPEM	Institutes of Physics and Engineering in Medicine
ISO	International Organization for Standardization
MDD	Medical Devices Directive
MDDS	Medical Device Data System
MDR17	Medical Device Regulation 2017/745
MDSAP	Medical Device Single Audit Program
MEDDEV	European Commission guidance documentation
MHRA	Medicine & Healthcare Regulatory Agency
NB	Notified Body
NPL	National Physical Laboratory
OTS	Off The Shelf
OTSS	Off The Shelf Software
PACS	Picture Archiving and Communications System
PEMS	Programmable Electronic Medical Systems
PMA	Pre-Market Approval
PMN	Pre-Market Notification
QMS	Quality Management System
SaMD	Software as a Medical Device
SOUP	Software Of Unknown Provenance
TR	Technical Report (IEC)

9.2   HEALTH INFORMATION PRIVACY

9.2.1   Introduction to health information privacy

Individuals and organizations have responsibilities regarding the confidentiality, integrity and availability of the data they collect or process. The regulatory frameworks in the European Union and United States are introduced below and some guidelines and principles of best practice highlighted.

9.2.2   Regulation in the European Union

The European regulatory framework is the *GDPR* (General Data Protection Regulation). It applies to all organizations (world-wide) that collect and/or process personal identifiable information of EU/EEA citizens (plus citizens from Norway, Iceland and Liechtenstein). It also applies to all processing of personal identifiable information that take place in EU/EEA for all individuals, whatever their nationality or place of residence, that is, even for non-EU/EEA citizens. For further details see Recital 14 and Article 3 of GDPR. The regulation aims to strengthen the protection relating to processing of personal data. It is not specifically developed for health information but a generic regulation that covers all types of personal identifiable information. In GDPR, health-related information is considered as *Special Categories* of personal data that is afforded extra protection under the regulation[13].

A care provider (e.g., a hospital) has the right, according to GDPR, to process health related information needed to provide diagnosis and treatment for an individual (called *Data Subject* in the regulation). The care provider does not need to ask the individual (patient) for any type of formal consent. However, if the data is to be used for anything other than treating the patient then consent is needed unless the care provider can demonstrate that the processing meets any of the other lawfulness criteria or if the data is anonymized. The regulation also gives individuals' rights over their personal information, for example, rights to examine and obtain a copy of their personal data (Article 15), and to request corrections (Article 16). Note that the right to object to processing and to delete data does not apply to all types of data, for example, healthcare data needed for treatment of patients (Article 17(c)).

In the regulation the care provider is most often the *Controller* (the organization that decides what the data should be used for) and any supplier or third party to the care provider (that can access patient data) is seen as a *Processor* or *Sub-processor* (an organization that is processing the data on behalf of the Controller). There must be a so-called *Data Processor Agreement* (DPA) set up between the parties in which the Controller specifies, among other things, which data the Processor can process, for what purpose, and what security measures are needed. The regulations do not specify what security measures that are needed. It is up to each Controller to decide appropriate security measures depending on the sensitivity of the data. Healthcare data is seen as sensitive data by the regulation and many care providers handle large amounts of healthcare data and will need to have very good security measures in place.

The definition of a personal data breach in GDPR (Article 4, bullet 12) is *"a breach of security leading to the accidental or unlawful destruction, loss, alteration, unauthorised disclosure of, or access to, personal data transmitted, stored or otherwise processed"*. Therefore GDPR does not only cover the confidentiality aspect of the personal data, but also the integrity (correctness) and availability aspects of the data. As soon as the controller becomes aware that a personal data breach has occurred, the controller should notify the supervisory authority within 72 hours after having become aware of it, unless the controller is able to

---

[13]GDPR does not apply to data that is anonymized (Recital 26) or to data concerning deceased individuals (Recital 27).

demonstrate that the breach is unlikely to result in a risk to the rights and freedoms of the individuals (Recital 85). In case the personal data breach is likely to result in a high risk to the rights and freedoms of the individuals the controller should also communicate the data breach to the affected individuals as soon as reasonably feasible (Recital 86).

Most European countries also have local regulations regarding healthcare data and patient identifiable information that also must be followed. These regulations can, for example, state when data can (and cannot) be transferred between hospitals or between departments within the same hospital. However, this text does not consider any of these local regulations. In addition to this hospitals often have internal policy's and rules that the employees of that hospital must follow.

9.2.3 Regulation in the United States

In USA there is a regulatory framework for health-related information called *HIPAA* (Health Insurance Portability and Accountability Act). This is a federal privacy law to protect individuals' medical records and other personal health information and applies to health plans, health care clearinghouses, and those health care providers that conduct certain health care transactions electronically. The HIPAA *Privacy Rule* requires appropriate safeguards to protect the privacy of personal health information and sets limits and conditions on the uses and disclosures that may be made of such information without patient authorization. The Privacy Rule also gives patients rights over their health information, including rights to examine and obtain a copy of their health records, and to request corrections (i.e., like GDPR).

The Privacy Rule protects the privacy of individually identifiable health information, called *Protected Health Information* (PHI) in any form or media, whether electronic, paper or oral. The *Security Rule* protects a subset of information covered by the Privacy Rule, which means all individually identifiable health information a care provider (called a Covered Entity in HIPAA) creates, receives, maintains or transmits in electronic form. The Security Rule calls this information *"electronic protected health information"* (e-PHI). Note that the Security Rule does not apply to PHI transmitted orally or in writing.

The Security Rule requires care providers to maintain reasonable and appropriate administrative, technical and physical safeguards for protecting e-PHI. Specifically, covered entities must:

1. Ensure the confidentiality, integrity and availability of all e-PHI they create, receive, maintain or transmit;

2. Identify and protect against reasonably anticipated threats to the security or integrity of the information;

3. Protect against reasonably anticipated, impermissible uses or disclosures; and

4. Ensure compliance by their workforce.

The Security Rule defines "confidentiality" to mean that e-PHI is not available or disclosed to unauthorized persons. "Integrity" means that e-PHI is not altered or destroyed in an unauthorized manner. "Availability" means that e-PHI is accessible and usable on demand by an authorized person.

The Privacy Rule permits a covered entity to use and disclose protected health information for research purposes, without an individual's authorization, under certain circumstances (see details in HIPAA itself for this). In HIPAA the care provider (e.g., hospital) is called the *Covered Entity* and any supplier or third party to the care provider that can access e-PHI is called a *Business Associate*. There must be a so-called *Business Associate Agreement* (BAA) set up between the parties.

9.2.4 Types of information covered

Europe (GDPR): All types of data elements that by themselves, or in combination with other data elements, could identify an individual (Data Subject), are covered by the GDPR.

USA (HIPAA): HIPAA applies to "individually identifiable health information", including demographic data, that relates to:

- the individual's past, present or future physical or mental health or condition,
- the provision of health care to the individual, or
- the past, present or future payment for the provision of health care to the individual,

and that identifies the individual or for which there is a reasonable basis to believe it can be used to identify the individual.

9.2.5 Standards and guidelines covering health information privacy

Official GDPR guidelines are provided by the European Data Protection Board (EDPB)[64]. Another good place to find GDPR guidelines in English is the UK Information Commissioners Office (ICO)[65]. A good source of HIPAA related security standards and guidelines is HIMSS (Healthcare Information and Management Systems Society)[66]. NIST (National Institute of Standards and Technology) have issued (September 2019) a draft guidance called Securing Picture Archiving and Communication systems[14]. The European Commission have issued (January 2020) a guidance on cybersecurity for Medical devices[15] to provide manufacturers with guidance on how to fulfil all relevant essential requirements of Annex I to the MDR17 and IVDR17 (In Vitro Diagnostic Medical Devices Regulation 2017/746).

9.2.6 Consequences of violations

Breach of GDPR and/or HIPAA could result in fines (penalties). Fines are primarily based on the number of individuals affected and the sensitivity of the information, but also if the organization have done its best to follow the regulation or if they deliberately have neglected the regulation as well as how co-operative they are with the authorities during and following an incident. Minor violations might only result in a warning letter and instructions to improve. Apart from having to pay fines to the authorities an organization might also be required to pay damages to any person who has suffered material or non-material damage as a result of an infringement of the regulation. (Note, there's a website which contains a list of fines that data protection authorities within the EU have imposed under the GDPR[16]. The US Department of Health and Human Services (HHS) also publishes a list of all PHI breaches where 500 or more individuals have been affected[17]).

9.2.7 Privacy best practices

Defined purpose and legal grounds
There must be a defined purpose for the processing that also states the legal grounds for the the data processing. Legal grounds (lawfulness) could be, for example, that the data

---

[14]https://www.nccoe.nist.gov/projects/use-cases/health-it/pacs
[15]https://ec.europa.eu/docsroom/documents/38941
[16]https://www.enforcementtracker.com/
[17]https://ocrportal.hhs.gov/ocr/breach/breach_report.jsf

use is needed for diagnosis and treatment, or based on consent from the individuals, or used to fulfil legal obligations, or for public interest, or for the purpose of legitimate interest pursued by the care provider. This is an area where legal advice often is needed.

Defined retention time

The retention time for the data must be defined. Personal identifiable data must not be kept longer than needed and there should be a process (routine) to delete data when no longer needed. This is also applicable for the log files if they contain personal data. There might also be legal requirements contained in local (national) patient data regulation that stipulate how long different types of health related data must be retained.

Data Protection Impact Assessment (DPIA)

DPIA is a GDPR concept for conducting a structured and documented impact assessment, but it is also useful in a HIPAA context. A DPIA is only required when the processing is "likely to result in a high risk to the rights and freedoms of the individuals". The DPIA should be carried out early in a project. To decide if a DPIA is needed, and how to conduct one, please refer to the EU guidelines on DPIA[67].

Anonymization (de-identification)

Data that is anonymized does not fall under any privacy regulations and there are therefore no privacy restrictions on what can be done with it. Consent is not required to anonymize data, or to use anonymized data, but the anonymization should be performed by an organization that has the right to process the data. According to GDPR, personal data is anonymized when it is not possible to identify an individual considering all the means reasonably likely to be used including objective factors, such as the costs of and the amount of time required for identification, as well as the available technology at the time of the processing and technological developments (for details see Recital 26 in GDPR). According to HIPAA personal health information is de-identified when there is no reasonable basis to believe that the information can be used to identify an individual (for details see paragraph 164.514 in HIPAA).

Example: To anonymize a radiology image, all data elements that could directly identify the individual, for example, personal identification number, social security number (or similar unique identifiers) must be removed. Data elements that indirectly and/or in combination with other data could identify the individual must also be considered, for example, name, date of birth, address, relatives, healthcare facility, date of examination, age, gender. If the image contains things like tattoos, scars, birth marks or the face of the individual (e.g., in a CT scan of a head), that data must also be "removed" or "blurred" if it could be used to identify the individual.

Anonymization is a complex area as there often is a level of probability involved and previous anonymization may need to be reviewed as the technology advances. The following three references provides more details and useful guides:

- Anonymization Code of Practice—by the ICO in UK [68].

- Guide for anonymization of research data—by the Finnish Social Science Data Archive (FSD) [69] .

- AIDA Data Sharing Policy. AIDA (Analytic Imaging Diagnostics Arena) is a Swedish arena for research and innovation on artificial intelligence for medical image analysis that has created a policy on what they think is good enough anonymization for medical images to be used in research and for publication in research reports [70].

Pseudonymization

Pseudonymization is the separation of data from direct identifiers so that linkage to an identity is not possible without additional information that is held separately (a key). For example, if radiology images are sent for external review to another healthcare facility, that receiving healthcare facility might not need to know the identity of the patients. In that case, data elements that could identify the individual can be removed (unless they are needed for the review), and instead a reference number can be assigned that can only be linked to a patient identity by the sending healthcare facility. Using pseudonymization reduces the risk of a confidentiality breach but will not make the data anonymized. The regulations still apply to pseudonymized data sets, but it is a good protection mechanism.

Minimize and limit

To reduce the sensitivity of the data one should always minimize the number of personal identifiable data elements, i.e do not collect more personal data elements than necessary for the purpose the data is collected.

Hide

Personal data that the user does not need to see should be hidden from the user. For example, bank account details might be needed for billing purposes and an address or phone number might be needed to contact the individual, but none of that information is needed for the radiologists to decide a medical diagnosis and should be hidden from the radiologist.

Rights of the individuals

The system must make it possible to meet the rights of the individual, for example, the right to receive a copy of their personal data, and the right to request corrections. This does not mean that it needs to be easy to perform these activities (it is acceptable to require manual database queries to be performed by a database expert) but it should be possible.

9.2.8   Security best practices

Protection by default

The default settings should always be set for maximum protection. If less protection is needed, then that should be a manual process to remove the not-needed protection mechanisms. Example of good security practises are: always apply latest security patches on the operating system, take regular backups (and test that the backup works), have up to date anti-virus software installed, encrypt all data in transit and (when possible) encrypt data at rest. It is also important that only the staff that need access to the data have access and no one else. This put emphasis on authorization and authentication, such as role based and or department/institution based access, as well as strong password policies with two factor authentication etc.

Separate

If sensitive data is separated from non-sensitive data it is possible to put extra protection on the sensitive data and less protection on the non-sensitive data, for example, encrypt the sensitive data but not the non-sensitive data.

Audit logging

All access to patient data should be logged. The logs should state who accessed the data and when. This means that every user (including system administrators and IT-staff) should use personal accounts and not shared accounts. Log files should also be protected from deletion and manipulation. A good practice is to automatically copy log files from the system to a separate log file system that only has read access. Logs should then also be checked, at specified intervals, for abnormal activities.

9.2.9 Conclusion regarding health information privacy

Even if there are differences between GDPR and HIPAA (e.g., GDPR covers all types of personal data, while HIPAA only covers healthcare data) the basic concept of protecting the confidentiality, integrity and availability of the data is the same. So, if a healthcare IT-system is designed and built for one of the regulations it will most likely also fit the other regulation regarding requirements for confidentiality, integrity and availability.

To be noted is that the integrity and availability aspect go hand in hand with safety requirements in a healthcare context, while the confidentiality aspect does not. It can sometimes be tricky to balance confidentiality requirements with safety requirements: too much protection could hinder an efficient clinical workflow. Care must be taken when designing the solution and "breaking the glass" functionality could be one option to solve that dilemma. That is, normal protocols and access privileges could be bypassed in an emergency, when it is critical for a user to access certain data.

List of new abbreviations for the second section of this chapter:	
AIDA	Analytic Imaging Diagnostics Arena
BAA	Business Associate Agreement
DPA	Data Processor Agreement
DPIA	Data Protection Impact Assessment
EDPB	European Data Protection Board
FSD	Finnish Social Science data archive
GDPR	General Data Protection Regulation
HHS	U.S. Department of Health and Human Services
HIMSS	Healthcare Information and Management Systems Society
HIPAA	Health Insurance Portability and Accountability Act
ICO	Information Commissioners Office
IVDR17	In Vitro Diagnostic Medical Devices Regulation 2017/746
NIST	National Institute of Standards and Technology
PHI	Protected Health Information

II

Problem-solving: examples from the trenches

## CHAPTER 10

Applying good software development processes in practice

Tanya Kairn

Cancer Care Services, Royal Brisbane and Women's Hospital, Herston, Queensland, Australia

Science and Engineering Faculty, Queensland University of Technology, Brisbane, Queensland, Australia

CONTENTS

T HIS chapter demonstrates how processes of good software development and validation provide opportunities for software to be efficiently focused on intended outcomes, for errors to be detected early and for users to thoroughly understand the code. Further, it also allows for revisions to be made and verified easily by individuals not involved in the initial code writing process, so that the usefulness of the software can be extended and the associated risks can be reduced.

10.1 INTRODUCTION

The key to adopting a good software development process, for any software development project, is the identification and use of a suitable software validation framework. While software *verification*, or the testing and de-bugging of each new element of the code, is an inherent part of software development, software *validation* is an additional, overlying process, which can require substantial additional planning, documentation and time. Software validation involves "confirmation by provision of objective evidence that software specifications conform to user needs and intended uses, and that the particular requirements implemented through software can be consistently fulfilled"[71]. While any software can function correctly, properly validated software has the additional advantages of being demonstrably complete, useful and robust.

Comprehensive software validation involves developing the processes and documentation needed to demonstrate quality management, which is necessary in order to certify or register software as a medical device. Depending on the features and intended use of the software as well as the jurisdiction in which it will be used, certification or registration as a medical device may be required in order for software to be sold, distributed, shared across national borders or shared between different institutions within one nation[71, 72, 73]. However, even if the software is developed in-house, for use within one institution, the software development process benefits from being managed under an appropriate quality management system, and this is even a requirement under the new EU Medical Device Regulation for Medical Device Software.

More than 15 years ago, the United States' Food and Drug Administration (FDA) published an influential report[71] providing guidance on software validation that, although intended to apply to software related to regulated medical devices, was based on established software validation principles and remains applicable to software development projects today, even when the software is not classed as a medical device. Two important aspects of the FDA's guidance make this framework especially useful for radiology software projects. Firstly, the FDA's guidance takes a risk-management approach to software development, emphasizing that the detail and extent of the planning, completion and documentation of software validation tasks should be commensurate with the complexity of the code and the risks associated with its use[71]. This emphasis on risk management is also apparent in the latest International Electrotechnical Commission (IEC) international standard on medical device software[72] and conforms with a broad trend towards using risk assessments, statistical process control and failure mode effects analysis to identify key quality assurance tests and manage the broadening workload of medical physicists[74, 75].

Secondly, the FDA's guidance is founded on an understanding of the particular issues that affect software relating to medical systems. Medical device software may be used for a range of diagnostic, therapeutic or analytical purposes, in multidisciplinary environments where users have different specializations and different levels of familiarity with the design, use or evaluation of software systems. Evaluation of medical device software should extend beyond the software's compatibility with local systems and effects on local users, to include the software's effects on patients as well as potentially remote individuals and systems, including referrers, consultants and databases.

This chapter uses the FDA's guidance as a framework to demonstrate the process of software validation, with reference to the more-recent IEC standard, using the example of a project that involved the development of in-house software for analysing radiochromic film measurement data.

10.2 THE TRENCH IN QUESTION: RADIOCHROMIC FILM DOSIMETRY

10.2.1 General features and applications

Radiochromic film is a high-resolution, two-dimensional dosimetry medium, that has been increasingly used in radiological applications since the early 2000s[76]. Radiochromic film is self-developing; exposure to radiation produces a color change that is immediately observable without chemical processing and increases non-linearly with dose, which can provide high-resolution two-dimensional measurements of radiation dose. Although originally designed and widely used as a dosimeter for radiation oncology[76, 77, 78, 79, 80], radiochromic film has also been used to measure and monitor imaging dose delivered to staff[81] and patients[82, 83, 84] during radiological procedures. The careful use of radiochromic film can allow accurate measurements of radiation dose to be correlated with records of radiation-induced skin toxicity[85] and estimates of radio-carcinogenesis risk[83].

The major differences between the radiation sources used for therapeutic and diagnostic procedures that affect the use of radiochromic film are the lower energies of diagnostic radiology sources (requiring radiochromic film to be calibrated using radiation sources that have the same or similar energies as the sources that are to be measured[86, 87]) and the lower doses that are delivered during diagnostic procedures[88] (requiring the specific type of radiochromic film used to be carefully selected with reference to advertised sensitivity ranges[88, 89]).

Radiochromic film has the potential to be a valuable dosimetry tool for many radiology departments, if the challenges associated with film calibration and analysis can be resolved. The development of in-house film analysis software provides an immediate solution to some of those challenges.

10.2.2 Film calibration and analysis

The process of calibrating and analysing radiochromic film dosimetry measurements is straightforward, but repetitive: an obvious candidate for automation. Calibration of radiochromic film for dosimetry generally involves irradiating small pieces of film to different known doses, digitizing (scanning) the film and then identifying a mean pixel value in the image of each piece of film. A calibration relationship can then be derived between the resulting pixel values and the corresponding delivered doses. To use the derived calibration relationship to convert the pixel values from a scan of a separate measurement film (e.g., a film placed on the patient's skin during an interventional radiology procedure) into an array of dose values, all scanning and analysis steps that were performed for the calibration films must also be performed for the measurement film. To minimize the repetition of manual analysis steps, a small number of commercial film dosimetry packages are available, although they can be costly and have limited adaptability. In particular, current systems do not allow users to perform steps that are especially important when measuring low doses from kilovoltage photon sources, including image averaging[89], application of scanner output corrections[89] or use of a non-standard calibration to minimize energy-dependence[86, 90].

In this environment, the attraction of developing in-house software is clear. Well-designed in-house software packages have the obvious benefit of fitting the needs, environments and workflows of the departments where they are developed[91].

10.2.3 Why treat this software as a medical device?

Categorization as a "medical device" affects the approvals needed to sell a particular piece of software into substantial commercial markets, including the European Union and the United States of America. While software written using MATLAB can be sold by individual developers and institutions, provided they have an appropriate agreement with Mathworks (e.g., sale of code may be covered by a commercial MATLAB license but not by a student MATLAB license), this chapter focuses on the common scenario where in-house code is developed for use within one institution.

Treating in-house software as a medical device provides a thorough framework, motivation and justification for software validation. In particular, treatment of software as a medical device requires verification steps that assess and take account of the harm can affect the patient, due to errors in the software[71, 72]. The FDA's guidance explains that comprehensive software validation "can increase the system's usability and reliability, leading to: decreased failure rates, fewer recalls and corrective actions, and less risk to patients and users' while also making it easier and less costly to reliably modify software and revalidate software changes"[71]. To be clear, it is not necessary for film dosimetry software (or any software) to actually *be* a medical device to use the advice provided in this chapter. The fact that the software is written by medical physicists, for use in radiology settings, where it has the potential to affect patients (even indirectly) means that validating the software as though it is a medical device is a useful way to manage and limit the risks associated with the use and future modification of the software.

10.3 AN IN-HOUSE SOFTWARE VALIDATION CHECKLIST

In-house medical physics software applications are often relatively small and uncomplicated, and are isolated to a small number of individuals, locations and tasks. Depending on their purpose and use, such software applications can nonetheless (directly or indirectly) affect patients and potentially cause harm. It is therefore important to use a risk-based analysis to identify the software validation steps that are specifically required for each project. The FDA's guidance notes that "For very low-risk applications, certain [validation] tasks may not be needed at all. However, the software developer should at least consider each of these tasks and should define and document which tasks are or are not appropriate for their specific application"[71].

For example, Figure 10.1 shows a proposed checklist for radiology physics software validation, which applies the key recommendations of the FDA's guidance without making software validation an insurmountable obstacle to software development. This checklist was generated by using a risk-based assessment to condense the FDA's guidance into a brief list of the key software validation elements that were required for the film analysis software development project. The specific items on the checklist are described in Sections 10.4, 10.5 and 10.6.

Validation task	1 Purpose defined	2 Stakeholders identified	3 Stakeholder requirements identified	4 Software design specification completed	5 Risk assessment completed	6 Required validation steps defined, based on specification and risk assessment	7 Software development plan completed, including software verification procedures	8 Independent design review completed	9 Coding completed, with coding and de-bugging activities logged	10 Code comprehensively commented	11 Developer testing completed and reported	12 User guide drafted	13 User acceptance testing completed and reported	14 Final report on software development completed, including validation against specification	15 Final report accepted by stakeholders
Date completed															
Location of documentation															

Figure 10.1: Checklist of software validation steps, for film dosimetry software development project.

10.4 BEFORE WRITING THE CODE

10.4.1 Defining the purpose

The software must have a clearly defined purpose, before development starts. Depending on the scale and complexity of the project, this definition may be in the form of a written description or proposal, a flow chart, a one-sentence summary or simply a name for the software. The purpose should be documented in writing. A draft definition of the proposed software's purpose can be used to communicate the aims of the software to potential users early in the project, so they can provide feedback and influence the final defined purpose of the software.

For example, the initial draft purpose of the film dosimetry software was, "To automate the process of deriving a calibration relationship using mean pixel values from TIFF images of multiple calibration films". Asking for feedback from potential users (physicists involved in the local film dosimetry programme) led to a substantial broadening of the purpose, to include calculating and reporting dose measurement results, to facilitate use by the entire local medical physics group. This change to the intended use of the software was associated with an increase in the level of risk arising from its use, requiring a more thorough software validation process.

10.4.2 Identifying the users and their requirements

One of the goals of the software validation process is to produce software that demonstrably conforms to user needs[71]. For example, the IEC's international standard on medical device software requires a requirements specification even for low-risk software[72]. It is therefore necessary to identify potential users and their requirements.

For the film dosimetry software the purpose and requirements of the code and the initial design specification were defined through discussion and consideration of the needs of stakeholders including the medical physicists who would use the code as well as doctors and other clinical staff who would require the film dosimetry process to produce clear and accurate results.

10.4.3 The risk assessment

The risk assessment allows the risks of using the software to be identified and mitigated, and allows the extent of required validation to be established and justified. The results of the risk assessment, and in particular the identified impacts that the software may have on patients, can determine the level of documentation, oversight and review required for certification of the software[72, 73]. The risk assessment should be completed before drafting the specification and the validation plan, so that the project planning steps can be completed with a thorough understanding of the risks expected when using the software. The risk assessment then informs both the writing and the testing of the code.

The risk assessment for the film dosimetry software followed a conventional process of considering hazards and vulnerabilities that might lead to risks, assessing the severity and likelihood of possible negative effects, identifying changes to software, documentation, validation or other processes that could mitigate those risks, and then re-assessing the risks assuming that mitigation was completed as planned. Broadly, hazards were identified by approaching the problem from two directions. Firstly, by considering the possible consequences of errors in input, output and calculation, which was made comparatively straightforward by the simplicity of the software. Secondly, by considering the known sources of error and uncertainty affecting film dosimetry measurements[92], and identifying how those errors or uncertainties could be produced or exacerbated by the use of the software.

Identified hazards included dosimetric errors due to user/software errors in ROI placement, image path data entry and interpretation, or calibration data entry and interpretation, as well as geometric errors due to user/software errors in image resolution data entry and interpretation. Undetected user errors were expected to affect individual measurements, whereas undetected software errors could affect large numbers of measurements. The risk assessment further identified the need to set the default resolution of the dose plane output to match the default resolution of the local film scanner and to provide users with simple, straightforward instructions on how best to use the software. The risk assessment also led to the drafting of a list of required validation steps, developer tests and user acceptance tests to be completed before the software could be released for use. The risk mitigation steps identified via the risk assessment process reduced the risks associated with all hazards down to low or negligible. It is unlikely that all of these risks could have been considered and mitigated without completing a deliberate and documented risk assessment process.

10.4.4 The software design specification

The software design specification is the key document at the center of the software validation process. The software design specification provides a logical description of what the software should do and how it should do it, in order to fulfill its purpose. The specification guides the developer to remain within the agreed requirements of the project and helps prevent "scope creep". The specification is also useful as a means for the developer to communicate the purpose and intended content of the software to others. A well-written specification can help justify staff and resource allocations or can be used to promote the work of the project team. Ultimately, however, the specification is most valuable as the basis upon which the software can be evaluated, to establish that the software meets the agreed end-user needs.

The design specification may be divided into separate "functional specification" and "technical specification" components, though combining both into one document is not unusual, for in-house clinical software projects[91]. As the specification provides a description of the planned features of the software against which the function and completeness of the finished code will be validated, the length and detail required in the specification depends on the size and complexity of the software development project and the outcomes of the risk assessment. The risk assessment should be re-visited and re-evaluated throughout the process of developing the specification.

For a large, complex, high-risk project (e.g., where use of the software has the potential to directly cause harm to patients), the requirements of the software design specification, and the whole validation process, are correspondingly large. The specification should provide details on how the software will work and how users will interact with it, including data structures, memory requirements, and user interface design[91]. The drafting of the specification may itself be a substantial project, requiring collaboration between individuals and institutions with different specializations.

For a small, local project, where possible unwanted effects on patients are minor and indirect, such as developing in-house film dosimetry code, the initial design specification may be comparatively brief. For the film dosimetry code discussed in this chapter, the design specification was written as a simple point-form list. The list consisted of 20 brief points that summarized inputs, analysis steps and outputs, plus a brief list of features that were desirable but out of scope. Aspects of the local film dosimetry method[92] were deliberately included in the specification, in order to help focus the software development project on one particular film analysis technique and minimize the potential for the software to become over-complicated and unwieldy.

10.4.5   The development plan

In addition to contributing to software documentation and validation[71], the development plan can be a useful project management tool. The software development plan should specify the "scope, approach, resources, schedules and the types and extent of activities, tasks and work items" involved in developing the software[71] and the IEC international standard requires a development plan to be drafted with specific content determined by hazard level, even for low-risk medical software[72].

The film dosimetry software project was small and did not have its own project management documentation (proposal, plan, etc), and so the software development plan needed to be drafted from scratch. Fortunately, the software design was so simple that the software development plan was briefly summarized by listing the development steps from selecting an appropriate programming language, to writing and debugging specific sections of the code, to testing, reporting, release and maintenance.

10.4.6   The validation, review and approval plan

The software validation plan defines and governs the software validation process, specifying the type and extent of validation activities that will be undertaken, while also providing relevant details regarding when, how and by who the validation steps will be completed. The FDA's guidance acknowledges that "the selection of validation activities, tasks and work items should be commensurate with the complexity of the software design and the risk associated with the use of the software for the specified intended use"[71]. Basic verification activities may be sufficient for lower risk software systems, additional validation activities should be added to cover any increases in risk, and "validation documentation should be sufficient to demonstrate that all software validation plans and procedures have been completed successfully"[71].

The software validation plan for the film dosimetry code is reproduced here, with asterisks indicating the steps that were modified as a result of the risk assessment:

1. Independent review of specification and plan

   (a) Record review date, reviewer name, outcome of review and follow-up, in software development documentation.

2. Ongoing verification testing and de-bugging during code writing.

   (a) Complete at least one independent code review, during development.

(b) Log software development progress and verification.

3. Comprehensive commenting of final code.

   (a) Retain a listing of the code before and after commenting.

4. *Developer testing of final code.

   (a) Use a sufficiently broad selection of test scenarios for structural testing, including execution and repetition of all loops.

   (b) Function test the performance of the software when handling expected and unexpected inputs (test the accuracy of ROI data and dose arrays against results from manual methods and test response to invalid input).

   (c) Document and report the developer testing of the code, recording the specific inputs that were used at each stage.

5. *Drafting of user guide.

   (a) Prepare a detailed user guide, emphasizing the need to use the recommended naming convention and perform a "sanity check" measurement (an irradiation of a measurement film using a known dose pattern) at the same time as each clinical measurement.

   (b) Include an explicit recommendation that any changes to the software must be thoroughly tested and documented, before the revised software is released for use.

6. *User acceptance testing.

   (a) Users should establish that software is correctly installed and operational.

   (b) Users should be given the report and data from the developer testing stage, for use in independent acceptance testing.

   (c) Retain all feedback and reports from acceptance testing.

7. Final report on software development.

   (a) Summarize the development process and validation outcomes.

   (b) Evaluate the software against the design specification.

   (c) Provide the final report to all stakeholders.

   (d) Present software development, documentation and validation information to local physics group. Note any resulting discussion and follow-up.

Note that in this example, the validation plan was longer than both the development specification and the development plan. The difference in length is indicative of the relative time and effort required to complete the validation steps compared to simply writing the code, which is discussed further in Section 10.7.

10.5   WHILE WRITING THE CODE

10.5.1   Independent design review

Critical self-assessment of the software design is difficult and inadvisable. Independent evaluation is preferable, whenever possible, and especially for higher-risk systems. Independent review may be undertaken by "internal staff members that are not involved in a particular design or its implementation, but who have sufficient knowledge to evaluate the project and conduct the verification and validation activities"[71].

During the development of the film dosimetry code, an internal reviewer (film dosimetry expert) was invited to examine the project documentation described in Sections 10.4.1 to 10.4.6 and an external reviewer (programming expert) was invited to examine the code. As a result of those reviews, minor changes were made that improved both the clarity of the documentation and the efficiency of the code.

10.5.2   Coding, de-bugging, testing and logging

At the code writing stage, the software validation enterprise switches from a focus on process management to a focus on verification (as distinguished in Section 10.1). For low-risk in-house software systems, most of the verification steps required to demonstrate software validation are probably covered by the de-bugging, testing and version control system that should already be part of routine software development practice. The validation process is completed by formal logging and reporting of those software development and testing steps, as a reference for future users or developers of the software, or for communication to external individuals or organizations, as evidence that the verification steps were completed. Testing should involve the production of input data or unit testing code that results in known outputs (including expected inputs, unusual inputs and erroneous inputs). These data may subsequently be used for systematic developer testing, after the code is complete, and shared with users, for the purpose of acceptance testing.

As local departmental policy prevented the use of an external version control system, each step in the process of writing, de-bugging and testing of the film dosimetry code was noted daily in a simple electronic log. This logging of the code writing process required minimal time and effort and became useful as a day-to-day workload tracking and management system.

10.5.3   Commenting

The FDA's guidance recommends that code should be clear, straightforward and well-commented[71]. This results in software that can be understood and more-safely updated and modified by the developer and others, after completion of the original software development project. While writing the code, the meaning and intention of each section are obvious to the developer and written explanations may seem unnecessary, so it is easy to start with good intentions and end up with poorly commented code. One way to handle this is to systematically review the code after completion but before release, deliberately adding comments throughout. When this final review of the film dosimetry code was completed, the number of lines of comments increased from 42 to 104. Many of the comments seemed excessive and condescending at the time, but proved more useful as time passed.

10.6 AFTER WRITING THE CODE

10.6.1 Developer testing

In addition to the documented software verification that is completed during development, a systematic process of re-testing the finished code should be completed, before starting the software release process. The rationale for testing should be derived from the software specification and validation plan, with reference to the risk assessment.

For the film dosimetry software, developer testing utilized a set of sample films and processes that allowed testing to cover every line of code and every path through the code (structure or "white box" testing[71]), while also testing the performance of the software in a broad range of clinical scenarios (functional or "black box" testing[71]). Examples of developer testing results are illustrated in Figure 10.2. Many of these tests were completed during development, but all were formally repeated and documented after code completion. These tests also helped refine the user acceptance procedure.

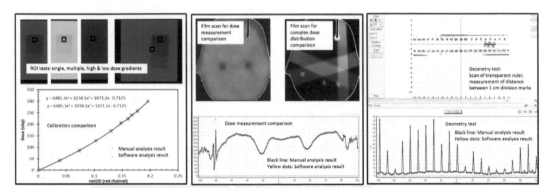

Figure 10.2: Developer testing results from film dosimetry software. Testing dose measurements against results of manual analysis: ROI values and calibration (*left*) and dose planes (standard and high-dose) (*center*). Testing dose plane geometry against physical reference: TIFF image of 1 cm markings on a 30 cm ruler measured at 1 cm separation in resulting dose image (*right*).

10.6.2 User guide and supporting documentation

User guides (software manuals) and user training are not explicitly addressed in the FDA's guidance document[71] although written instructions are specifically required by other jurisdictions. The results of the risk assessment may influence the drafting of a user guide or the composition of error and help messages within the software.

The user guide for the film dosimetry software provided step-by-step instructions, with illustrations, examples and references, as well as a "vulnerabilities, warnings and disclaimers" section, with a list of situations (identified via developer testing) where the code would halt due to user error and a list of situations (identified via the risk assessment) where the code would produce inaccurate results due to user error.

10.6.3 User acceptance testing

User acceptance testing generally includes software installation, inspection and testing at the user site, with appropriate documentation to demonstrate that proper installation has been completed. User site testing involves "actual or simulated use of the software ... within the context in which it is intended to function"[71], including the investigation and resolution or errors and the supply of a final test report[71].

For the film dosimetry software, which was to be used in-house at the same institution at which it was developed, an acceptance testing programme was completed by members of the physics group who were not involved in software development. Testing aimed to identify whether appropriately experienced medical physicists could, by simply following the written user guide, successfully install and run the software without further guidance, and to establish whether independent users would achieve the same results when repeating the "developer testing" tests (Section 10.6.1).

10.6.4   Validation against specification and final report

The FDA's guidance recommends that the software's end users should be provided with documentation that includes the defined user requirements, the validation protocol used, software acceptance criteria, test cases and results, and "a validation summary that objectively confirms that the software is validated for its intended use"[71]. This information can be efficiently provided in the form of a final report. The final report on the software development project should verify that the requirements set out in the planning documents have been adhered to: that the software development process followed the development plan, that the validation tests complied with the validation plan, and especially that the completed software complies with the design specification. By confirming and documenting this adherence to the planning documents, the final report demonstrates that risks have been managed and defined user needs have been met. The report then acts as evidence supporting the release of the software for clinical use, and as a reference guiding that use. For a well-planned and carefully validated low-risk software development project, the final report may be relatively brief. For example, the film dosimetry software development project produced a final report that contained just a short introduction and the completed check-list shown in Figure 10.1, including links to the network locations of all supporting documents.

10.6.5   Maintenance and software changes

Software validation and verification activities should be conducted throughout the entire software life cycle. In particular, the validation status of the software needs to be re-established after any change is made to the software after release. Sufficient testing should be undertaken to demonstrate that the altered parts of the software are operating correctly and that the unaltered parts of the software have not been adversely affected by the change. This may involve repeating some or all of the tests used for developer testing and acceptance testing and documenting the results. As during the initial software development project, however, testing alone is not sufficient to properly validate the updated software. Further documentation steps should be completed to record the purpose of the change, any increased or decreased risks, and the proposed methods by which the change is to be implemented and tested. In other words, the software validation steps completed when the code was initially developed should be reviewed and documentation should be updated as necessary.

10.7   SUMMARY OF VALIDATION PROCESS AND OUTCOMES

Generally speaking, the steps involved in writing low-risk in-house software for specific, local medical physics purposes can be summarized as: 1. Need the code; 2. Write the code; and 3. Use the code. For the film dosimetry software discussed in this chapter, completing those three steps took approximately three days of physicist-time and less than one week of calendar-time, but would have resulted in software that could only be used by the developer (unless a substantial amount of time was spent training each new physicist who wanted to use it), in very limited circumstances (research only, not anything that could influence decisions relating to a patient's treatment).

Completing the additional steps involved in using good software development processes including documented software validation steps, as listed in Figure 10.1, took eight days of physicist-time and two months of calendar-time, with the largest delays occurring while waiting for busy medical physicists to provide reviews and complete acceptance tests. This software validation process produced software that: could be used by other local medical physicists, without training; met users' needs, beyond its initially proposed purpose; remained easy to modify and maintain, by the developer and others, for months and potentially years after the initial project was completed; and had been independently reviewed and thoroughly tested by the developer and the users. Most importantly, the validation process resulted in software for which the risks to users, the local institution and the patients had been demonstrably identified and mitigated, and the development processes (including planning, writing, testing and acceptance) had been fully documented. This allowed the local medical physics group to formally release the software for clinical use.

10.8   REGARDING CERTIFICATION

It should be understood that completion of the checklist provided in Figure 10.1, or even compliance with the FDA's *guidance*, cannot be regarded as synonymous with compliance with FDA *regulations*, the IEC standard or with the requirements of any other jurisdiction. Developers who intend to produce software at a standard where it may achieve formal regulatory approval should examine the appropriate documentation (including Chapter 9 of this book, and its references) and/or seek guidance directly from the relevant regulatory authority.

10.9   CONCLUSION

This chapter provides an example of a process of software validation that was completed with the goal of producing a demonstrably complete, useful and robust software tool for analysing film dosimetry measurements, that could be used with more confidence, by more medical physicists, with less risk, than could have been achieved without software validation. This chapter therefore contends that software validation is largely its own reward, and that software validation processes should be completed when developing any medical physics software, even if the intended use of the software is in-house, non-commercial and not a medical device. Further, this chapter should act as a useful example of how to use a good software development process to produce in-house software that is more complete, useful, robust and reliable than it would have been without systematic software validation.

Acknowledgements
The author acknowledges Scott Crowe, who assisted in developing the film dosimetry software, as well as Bess Sutherland, Mark West and Nigel Middlebrook, who provided advice, opinions and independent testing as part of the software validation process.

## CHAPTER 11

Automating quality control tests and evaluating ATCM in computed tomography

Patrik Nowik

Department of Clinical Science, Intervention and Technology, Karolinska Institutet, Stockholm, Sweden

Medical Radiation Physics and Nuclear Medicine, Karolinska University Hospital, Stockholm, Sweden

CONTENTS

I N this chapter MATLAB is used to analyze CT images of various test objects. Pixel values are analyzed through the application of region-of-interests and several methods for finding specific objects in images are shown. Further, the presented concepts are implemented in an automatic workflow to analyze a series of images and extract image quality data for constancy tests and characterization of the automatic tube current modulation (ATCM).

11.1 INTRODUCTION

A medical physicist working in radiology will come across image analysis tasks. The analysis often consists of placing a region-of-interest (ROI) over an image and applying a statistical analysis of the pixel values within the region, typically acquiring the mean and standard deviation. The most commonly used method, when dealing with CT images, is probably to perform the analysis directly on the CT scanner console. The drawback with this method is that it is performed slightly different each time, both regarding ROI size and ROI placement. The first step of improvement is to move away from analysis in the console to your own computer by extracting the DICOM images, either by image export in the console or by a network transfer to a DICOM server. With the images available, the analysis can be done according to the user's preference. Scripts with automatic analyses can be set up, allowing the implementation of automatic workflows for image quality and trend analysis (see also Chapter 5). It is also possible to compile .m-scripts into .exe-files for execution on computers that do not have MATLAB installed (see Chapter 8). This chapter introduces CT image analysis with ROIs and design of analysis workflows.

11.2 ANALYZING CT PHANTOM IMAGES

Circular ROIs are often used for analysis of CT phantom images. The function below is used to create a circular mask:

```
function outMask = circleMask(roiCenter, imageSize, radiusCentral)
% Inputs
% phantomCenter: center coordinates of the ROI
% imageSize: size of mask image (same as image size)
% radiusCentral: radius of the ROI
% Outputs
% outMask: mask with value of 1 within circular ROI (value 0 outside)

cx = roiCenter(1); % Center X-coordinate of wanted circle
cy = roiCenter(2); % Center Y-coordinate of wanted circle
ix = imageSize(1); % Number of columns in original image
iy = imageSize(2); % Number of rows in original image
[x,y] = meshgrid(-(cx-1):(ix-cx), -(cy-1):(iy-cy));
outMask = ((x.^2+y.^2) <= radiusCentral^2);
```

The pixels of the mask image take either a value of 1 (inside the ROI) or 0 (outside the ROI). The mask can be used to extract the desired pixel values for analysis. See the example below where a circular ROI with a radius of 10 pixels is put at the center of a CT image and the mean and standard deviation of the pixel values in the circle are calculated. Here, the CT image (**CTImage**) is an image slice stored as a 2D array.

```
imageSize = size(CTImage);
roiCenter = imageSize/2; % Example with ROI positioned at image center
radiusCentral = 10; % Example with an ROI with a radius of 10 pixels
outMask = circleMask(roiCenter, imageSize, radiusCentral);
pixelValuesROI = CTImage(outMask); % Extraction of pixel values in the ROI
meanROI = mean(pixelValuesROI);
stdROI = std(pixelValuesROI);
```

The in-built MATLAB function **regionprops** is often useful as part of phantom image analysis scripts. It can be used to find the phantom center and size (in pixels) or other features of the phantom. In the example below a binary image, generated based on a CT number threshold of -300, is analyzed by **regionprops** to get the coordinates of the phantom center (**phantomCenter**) and the phantom size (**phantomSize**):

```
phantomData = regionprops(CTImage>-300); % phantomData is a struct array
phantomCenter = phantomData(1).Centroid; % In this example there is only one region
phantomSize = mean(phantomData(1).BoundingBox(3:4));
```

The structure (or *struct*) returned by **regionprops** for each region has three properties by default: **Centroid**, **BoundingBox** and **Area**. The **Centroid** property defines the center-of-mass of the region. The **BoundingBox** property defines the smallest rectangle that contains the region; the first two elements of the field define the top-left corner, the second two the size in the two dimensions. In the code above, the size of the phantom is defined as the average of the rectangle's width in both dimensions. In this example, `regionprops` is simply passed a binary image (i.e., logical array) in which all the non-zero values are *connected* (i.e., no isolated islands). For images with disconnected regions exceeding the threshold (or in which non-binary images are input), properties will be output for more than one region. A method is then needed to find the region desired (i.e., the appropriate index for `phantomData`). For example, the largest identified area (given by the **Area** property) is probably the phantom itself. We see this idea in action later, in Section 11.4.2.

Below, a function `water()` is presented for image analysis of a water phantom. It is programmed to output the mean values and the standard deviations in the ROIs that are needed for estimation of noise, CT number and uniformity. A *struct* containing the necessary information (`waterInput`) is used as an input to the function and the output is another *struct* (`waterResults`). Note that in `waterInput`, the positions for the uniformity test (the distance from phantom border) and the ROI diameter for noise, CT number and uniformity tests, are specified as a proportion of the phantom size. The function starts by finding the phantom and calculating the positions for the ROIs that will be needed:

```
function waterResults = water(waterInput)
% Inputs
% waterInput.HU: CT image (2d array)
% waterinput.positionUniformity: ROI positioning for uniformity test
% waterInput.diameterCentral: ROI size for noise test
% waterInput.diameterUniformity: ROI size for CT number and uniformity test
% waterInput.instanceNumber: Image number

% Outputs
% waterResults: analysis data

HU = waterInput.HU;
positionUniformity = waterInput.positionUniformity;
diameterCentral = waterInput.diameterCentral;
diameterUniformity = waterInput.diameterUniformity;

phantomData = regionprops(HU>-300);
phantomCenter = phantomData(1).Centroid;
phantomSize = mean(phantomData(1).BoundingBox(3:4));

radiusCentral = diameterCentral/2*phantomSize;
radiusUniformity = diameterUniformity/2*phantomSize;
positionUniformity3 = phantomCenter+[(0.5-positionUniformity)*phantomSize 0];
positionUniformity6 = phantomCenter+[0 (0.5-positionUniformity)*phantomSize];
positionUniformity9 = phantomCenter-[(0.5-positionUniformity)*phantomSize 0];
positionUniformity12 = phantomCenter-[0 (0.5-positionUniformity)*phantomSize];
```

Using the position information, image masks are created and the outputs (mean and standard deviation) are calculated by applying those masks to the CT image. Figure 11.1 shows images of a water phantom with ROIs depicted that correspond to those used in a demonstration of the water function.

```
maskCentral = circleMask(phantomCenter, size(HU), radiusCentral);
maskUniformityC = circleMask(phantomCenter, size(HU), radiusUniformity);
maskUniformity3 = circleMask(positionUniformity3, size(HU), radiusUniformity);
maskUniformity6 = circleMask(positionUniformity6, size(HU), radiusUniformity);
maskUniformity9 = circleMask(positionUniformity9, size(HU), radiusUniformity);
```

(code continues on next page)

```
maskUniformity12 = circleMask(positionUniformity12, size(HU), radiusUniformity);

waterResults.phantomCenterX = phantomCenter(1);
waterResults.phantomCenterY = phantomCenter(2);
waterResults.meanCentral = mean(HU(maskCentral));
waterResults.meanUniformityC = mean(HU(maskUniformityC));
waterResults.meanUniformity3 = mean(HU(maskUniformity3));
waterResults.meanUniformity6 = mean(HU(maskUniformity6));
waterResults.meanUniformity9 = mean(HU(maskUniformity9));
waterResults.meanUniformity12 = mean(HU(maskUniformity12));
waterResults.stdCentral = std(HU(maskCentral));
waterResults.stdUniformityC = std(HU(maskUniformityC));
waterResults.stdUniformity3 = std(HU(maskUniformity3));
waterResults.stdUniformity6 = std(HU(maskUniformity6));
waterResults.stdUniformity9 = std(HU(maskUniformity9));
waterResults.stdUniformity12 = std(HU(maskUniformity12));
```

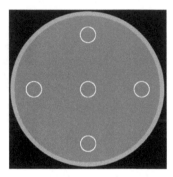

Figure 11.1: Demonstration of the ROI placement for measurements of noise (*left*), CT number of water (*middle*) and uniformity (*right*).

11.3 APPLICATIONS IN CONSTANCY TESTS

Quality Assurance (QA) is a part of quality management relating to ensuring that examinations at the clinic fulfil all quality related requirements. Within this concept, separate technical tests to ensure that the equipment is functioning as intended constitute Quality Control (QC) procedures. For CT scanners, the International Electrotechnical Commission (IEC) have published three main standards for CT equipment performance that forms the basis for manufacturing and QA programs: the IEC 61223-3-5[93], the IEC 61223-2-6[94] and the IEC 60601-2-44 Edition 3 Amendment 1[95]. These standards provide guidance on various tests, including frequency, methods and tolerances. A typical QA program of a CT scanner consists of periodical QCs that are performed annually. Often such annual QCs consist of tests that can be heavily correlated, for example, dose and noise. In CT, noise can be seen as a *Key Performance Indicator* (KPI) of a scanner[96]. A KPI is a readily measurable parameter that is affected by more fundamental parameters that determine scanner performance[3].

11.3.1 Extracting KPIs in a water phantom and compare to tolerances

A QA program called MonitorCT was developed which evaluates KPIs automatically[3]. The software was developed in MATLAB and the KPIs that are evaluated are: CT numbers,

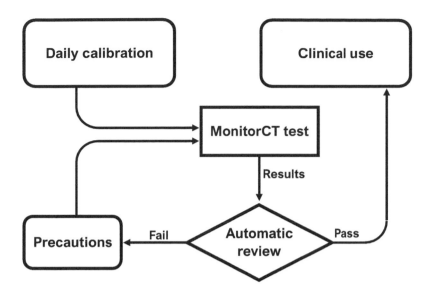

Figure 11.2: A schematic plan of the MonitorCT QA program. After the daily air calibration performed each morning, a QC test is performed. If the test has passed, the CT scanner is ready for clinical use. If the test has failed, precautions are taken.

noise, uniformity, homogeneity and positioning. The first important decision made was that it should be possible to extract all the selected KPIs using a vendor's QA phantom (a water phantom delivered with each CT scanner). This meant that the program would work with every CT scanner, since such a phantom is always available. The vendor's QA phantom includes a part consisting of plain water and in some cases other image quality testing parts. In developing MonitorCT, the focus was on using the water part, as it provided the possibility for the testing of KPIs sensitive to image quality and x-ray tube output (i.e., dose). Laser accuracy is still an area where separate tests are needed as the vendor's QA phantoms were not considered good enough for the job (this applies to all main CT vendors).

MonitorCT was developed such that the CT radiographers scan the vendor's QA phantom after the daily CT air calibration, using a predefined scan protocol. The scanners were set up so that the image series generated was automatically transferred to a DICOM-server, which started a compiled MATLAB program to perform the analysis of the KPIs (see schematic plan in Figure 11.2). A compiled MATLAB program can be started automatically from a DICOM server with the series path as an input, which is how the MonitorCT program was triggered. The main MonitorCT function was built around smaller functions that were started depending on the section of the phantom to be analyzed. Building a program that uses a main function that in turn invokes dedicated functions for solving specific problems is a recommended approach, as it makes the structure of the program clear and easier to work with.

The MonitorCT program was designed to connect to a database multiple times for various tasks, such as obtaining expected slice positions, tolerances and storing of the results. This provided additional flexibility and was convenient for trend analysis. Further, a web page was used for displaying the results of the tests. However, this chapter will not cover how to set up the database connections. Instead, a simplified version of the core MonitorCT program will be demonstrated. For information on how to set up a structured database with your project, see Chapter 5. The first thing that is done is a check that the program was started with a valid folder with DICOM images. The program will terminate if the folder does not contain CT images. If any non-CT-image files are found in the folder, they are omitted in the analysis. After this part of the program, the cell array `allPaths` contains the full paths to each CT-image in the input directory:

```matlab
function main(input)
% inputs
% input: the path to the DICOM files of the test


% Get all objects in folder to be analysed
allPaths = dirPlus(input); % Function available in File Exchange (by K Eaton)
% Check if objects are DICOM and if they can be analysed
go = zeros(numel(allPaths),1); % Filetype vector; only analyse if go is = 1
for i = 1:numel(allPaths)
    imagePath = allPaths{i}; % Image path to image i
    if isdicom(imagePath) % If DICOM-file, proceed
            meta = dicominfo(imagePath); % Get DICOM header
            modality = meta.Modality;
            width = meta.Width;
            if strcmp(modality,'CT') && width == 512 % Hardcoded image size
                go(i) = 1; % A CT image with 512x512 pixels: ok
            end
            if meta.SeriesNumber==999 || meta.SeriesNumber==997 % GE
                go(i) = 2; % A dose screen or dose report, not ok
            end
            if meta.SeriesNumber==501 || meta.SeriesNumber==502 % Siemens
                go(i) = 2; % A dose screen or dose report, not ok
            end
    end
end
if sum(go==0)
    disp('No DICOM files in folder')
    return % Exit the MonitorCT script
end
if sum(go==0)>0 % Remove non-DICOM objects
    allPaths(go==0)=[];
    go(go==0)=[];
end
if sum(go==2)>0 % Remove dose screens
    allPaths(go==2)=[];
    go(go==2)=[];
end
```

After the initial check, basic test information (such as test date) is collected and stored in a *struct* (`scannerData`) that could be saved together with the results:

```matlab
% Get basic series data
imagePath = allPaths{1};
meta = dicominfo(imagePath); % meta is the DICOM header of image number 1.
scannerData.date = meta.(dicomlookup('0008','0020'));
scannerData.time = meta.(dicomlookup('0008','0030'));
scannerData.institutionName = meta.(dicomlookup('0008','0080'));
scannerData.stationName = meta.(dicomlookup('0008','1010'));
scannerData.seriesDesc = meta.(dicomlookup('0008','103E'));
```

Before looping through the CT-images, the input data are defined for the analysis of noise and uniformity (ROI sizes and positions for input into the `water()` function). The slice number of the positioning part of the manufacturer's QA phantom is also defined (`positioningSlice`).

```matlab
% ROI placements relative to the phantom diameter
waterInput.positionUniformity = 0.15; % How close to phantom boarder ROI is placed
waterInput.diameterCentral = 0.4; % ROI size for noise test
waterInput.diameterUniformity = 0.1; % ROI size for CT number and uniformity tests
% Slice positions (all images treated as water if not having these image numbers)
positioningSlice = 1;
airSlice = 10;
```

The following code is the loop with the image analysis. For each image, the slice number is analyzed to decide which analysis script to start. The function `water()` was demonstrated in the beginning of this chapter. That function is the backbone of MonitorCT. It is possible to extract multiple KPIs, with noise being the most important, from CT images of the water part of a vendor's QA phantom. Each image is fed into an analysis function that gives an organized *struct* (`waterResults`) with the results. These results are stored row by row in a cell array (`allResults`). Tolerances of the KPIs can be used by including an analysis of `allResults`, but this is not shown in this example below:

```matlab
allResults={}; % Cell for storage of relevant data and results

for imageNumber = 1:numel(allPaths) % A loop through all images of the test
disp(['Image ' num2str(imageNumber) ' out off ' num2str(numel(allPaths))])
    imagePath = allPaths{imageNumber}; % Path to image to be analyzed
    meta = dicominfo(imagePath); % DICOM header of the image to be analyzed
    allResults(imageNumber,1,1:3) = {'StationName','StudyDate','StudyTime'};
    allResults(imageNumber,2,1:3) = {meta.StationName,meta.StudyDate,...
        meta.StudyTime};
    instanceNumber = meta.(dicomlookup('0020','0013')); % Image number
    HU = double(dicomread(imagePath))-1024; % Get image (CT images are rescaled)

    switch instanceNumber
        case positioningSlice % Analysis of an image of the positioning part
            imageType = 'positioning'; disp('Positioning')
            % Analysis for POSITION. Left out in this demonstration.
            allResults(imageNumber,1,4) = {'imageType'};
            allResults(imageNumber,2,4) = {imageType};
        case airSlice % Analysis of an image of air
            imageType = 'air'; disp('Air')
            % Analysis for AIR. Left out in this demonstration.
            allResults(imageNumber,1,4) = {'imageType'};
            allResults(imageNumber,2,4) = {imageType};
        otherwise % Analysis of an image of the water part
            imageType = 'water'; disp('Water')
            waterInput.HU = HU;
            waterInput.instanceNumber = instanceNumber;
            % Start analysis
            waterResults = water(waterInput); % The analysis function.
            % Store data and analysis for each analyzed image
            allResults(imageNumber,1,4) = {'imageType'};
            allResults(imageNumber,2,4) = {imageType};
            allResults(imageNumber,1,5:18) = fieldnames(waterResults)';
            allResults(imageNumber,2,5:18) = struct2cell(waterResults)';
    end
end
```

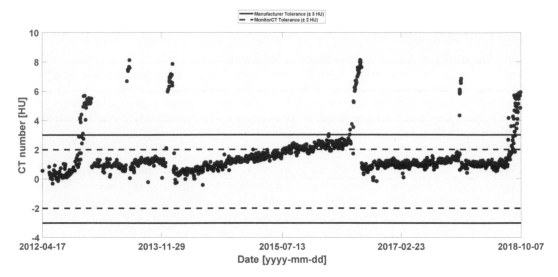

Figure 11.3: CT numbers of a CT scanner acquired from daily QC test over a period of six years. The vendor tolerances (3 HU) and MonitorCT tolerances (2 HU) around the baseline are indicated.

The cell with all the results is exported to a .csv-file in the end of the script using the MATLAB function `writecell`:

```
% Results to .csv
writecell(allResults,'results.csv')
```

Storing results as .csv-files give the possibility to open and analyze results fast, without the need of opening MATLAB. For example: the .csv-files could be saved in a folder shared through a cloud storage service (such as Dropbox or Onedrive) for easy access on other devices such as a mobile phone.

11.3.2   Data visualization (trends) and interpretation

At our institution it is uncommon that errors or problems are found at yearly QCs. However, with the introduction of the MonitorCT program, we have started to discover things of value to the clinics, as we have identified issues not previously seen. An example from a successful introduction of daily QCs is displayed in Figure 11.3 where CT numbers over a period of six years are displayed for a radiology clinic with high demands on CT number accuracy. In this example, CT number deviations were found at multiple occasions.

Finding such CT number deviations is crucial in radiology clinics with high demands on low contrast visibility. It is questionable if a conventional approach with an annual QC would be of much help to the clinic in the presented example. Timely identification of issues is critical, so that the vendor can be informed of problems before they impact the quality of clinical examinations. Based on our experiences, we have recommended daily QCs of all CT scanners at our institution.

11.4 APPLICATIONS IN AUTOMATIC TUBE CURRENT MODULATION

11.4.1 Automatic tube current modulation

One of the CT physicist's primary tasks at a radiology clinic is to support it with knowledge about the dosimetric consequences from adjustments of different scanner options. One of the least understood functions of CT scanners is probably the automatic tube current modulation (ATCM). It is used in the majority of CT scans and can make or break the image-quality of examination. The ATCM adapts the tube current of the x-ray tube during the CT scan acquisition in order to maintain image quality and radiation dose throughout individual CT scan volumes and between CT scans of differently sized patients. The ATCM works in the Z-direction (along the patient) and/or in the X/Y-direction (angular tube current adaptation during each rotation). The ATCM in the Z-direction is generally planned using the patient size extracted from the CT Localizer Radiograph (LR), while the ATCM in the X/Y-direction is planned either from the LR or from detector signals from previous rotations (online modulation). Beyond patient size, the ATCM uses the selected scan settings to further adapt the tube current. What settings and how they are used is vendor-, model- and software-dependent and therefore it is a challenge for CT physicists to keep track of specific implementations. Getting help from the vendors is not always an option, as our experience is that the local service representatives and product specialists may struggle to fully understand the ATCM as well.

At the time of writing there are no robust or comprehensive tests recommended for the ATCM function. There is also no standard phantom for the purpose of QC of ATCM or characterization of it.

11.4.2 Characterizing the ATCM

At our institution, we were in particular need of a way to characterize the ATCM function. We had CT scanners from four different manufacturers. Further, we saw that the ATCM differed between them and we did not understand precisely how. This project started with us developing a phantom that fulfilled our requirements for a good characterization of the ATCM. Our phantom was later on further developed and is now available commercially (ATCM-phantom, The Phantom Laboratory, Salem, New York, USA): it is shown in Figure 11.4. In the end, our work resulted in an extensive table of which parameters affect the ATCM and how[97].

Figure 11.4: An automatic tube current modulation phantom that can be used to characterize the ATCM function.

The ATCM phantom was scanned in CT scanners from the four different vendors using a standard scan protocol, changing one parameter at the time. This work resulted in multiple CT series to be analyzed and this was all done using MATLAB. The code to analyze each series of the phantom was built as a function (**getATCM**) that took the path (or paths) to the series as an input. The function delivered three vectors as output: the slice positions, the tube current per slice position and the image noise in each slice. A shortened version of the analysis script using the function is presented below. How these values can be plotted over a synthetic localizer radiograph (SLR) is also demonstrated. An SLR is a LR-like image created from a volume CT scan (generally a spiral scan), by averaging the volume over one of the dimensions. The analysis script starts with the creation of the SLR:

```
% Create synthetic localizer radiograph
[rawDicomVolume, ~, ~] = dicomreadVolume(seriesPath); % Read volume
dicomVolume = squeeze(rawDicomVolume); % Remove unused dimension
% Calculate SLR by calculating mean along 1st dim (ignoring nans)
syntheticLocalizerRadiograph = squeeze(nanmean(dicomVolume,1));
```

The script then calls the **getATCM** function:

```
[slicePositions, tubeCurrents, imageNoises] = getATCM(seriesPath);
```

This function loops through each image in the CT series, obtaining the slice position and tube current from the DICOM header as well as the image noise in each image:

```
function [slicePositions, tubeCurrents, imageNoises] = getATCM(seriesPath)
% Inputs
% seriesPath: path to the folder with the CT scan
% Outputs
% slicePositions: slice positions
% tubeCurrents: tube currents

fileNames = dirPlus(seriesPath); % function available in File Exchange (by K Eaton)
slicePosition = nan(size(fileNames,1),1);
tubeCurrent = nan(size(fileNames,1),1);
imageNoise = nan(size(fileNames,1),1);
for i=1:size(fileNames,1)
    imagePath = fileNames{i};
    HU = double(dicomread(imagePath))-1024; % Get image (CT images are rescaled)
    dicommeta = dicominfo(imagePath);
    slicePosition(i) = dicommeta.SliceLocation;
    tubeCurrent(i) = dicommeta.XRayTubeCurrent;
    imageNoise(i) = getATCMImageNoise(HU);
end
% Sort tube currents
[slicePositions,I] = sort(slicePosition);
tubeCurrents = tubeCurrent(I);
imageNoises = imageNoise(I);
```

The function for the calculation of the image noise (**getATCNImageNoise**) is presented below. It includes the previously discussed functions **circleMask** and **regionprops**:

```
function imageNoise = getATCMImageNoise(HU)
% Inputs
% HU: image
% Outputs
% imageNoise: image noise

diameterCentral = 0.4; % ROI 40 % of phantom diameter
phantomData = regionprops(HU>-300);
[~, maxAreaIndex] = max([phantomData.Area]);
if phantomData(maxAreaIndex).Centroid(2)>384
    imageNoise = nan; % If no phantom, set noise to nan
else
    phantomCenter = phantomData(maxAreaIndex).Centroid;
    phantomSize = mean(phantomData(maxAreaIndex).BoundingBox(3:4));
    radiusCentral = diameterCentral/2*phantomSize;
    maskNoise = circleMask(phantomCenter,size(HU),radiusCentral);
    imageNoise = std(HU(maskNoise));
end
```

11.4.3 Data visualization and interpretation

In the final code extract, we demonstrate how the tube currents can be plotted over the SLR. Image noise can easily be plotted by replacing `tubeCurrents` with `imageNoises`.

```
% Plot ATCM curves on top of synthetic localizer radiograph
image2plot = syntheticLocalizerRadiograph;

% Find range of slice positions for x-axis of image
slicePositionMin = min(slicePositions);
slicePositionMax = max(slicePositions);
slicePositionSteps = size(image2plot,2); % Image size in X-direction
imageXRange = linspace(slicePositionMin, slicePositionMax, slicePositionSteps);

% Find range of tube currents for y-axis of image
tubeCurrentMin = min(min(tubeCurrents))-50;
tubeCurrentMax = max(max(tubeCurrents))+50;
tubeCurrentSteps = size(image2plot, 1); % Image size in Y-direction
imageYRange = linspace(tubeCurrentMin, tubeCurrentMax, tubeCurrentSteps);

% Find pixel-value range for setting window settings
pixelRange = [min(min(image2plot)), max(max(image2plot))];

% Plot image first, followed by tube currents plotted over the image
imagesc(imageXRange, imageYRange, image2plot, [pixelRange(1)*1.2 pixelRange(2)])
colormap gray; hold on
plot(slicePositions, tubeCurrents, '.'); hold off % Plot mA on top
figureAxis = gca;
figureAxis.YDir = 'normal'; % Y-axis direction opposite when using imagesc
xlabel('Slice position [mm]')
ylabel('Tube current [mA]')
```

An example output is shown in Figure 11.5. The CT scan was performed using an ATCM function that aims at keeping the image noise constant, which the CT scanner clearly achieves. A way of starting the process of understanding the ATCM of your CT scanner is to scan a phantom and plot the used tube currents over the SLRs as in this example[1].

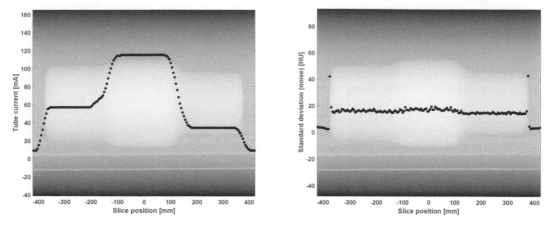

Figure 11.5: The tube current (left) and image noise (right) from a CT scan of a phantom using automatic tube current modulation.

11.5 CONCLUSIONS

QA programs like MonitorCT are likely to be increasingly used in the future. In many places the role of the CT physicist is changing from mainly performing manual testing, to being a more general support to optimization in the radiology clinic, investigating such things as ATCM. The most important practical take-home message from this chapter, however, is the structure of working with larger programs. It is easy to work with the program when it is broken down into small parts (functions) which have clear inputs and outputs. It is also easier to get an overview of the program, simplifying the discovery of bugs and weaknesses. Further, we should avoid repeating ourselves, writing the same code twice (or using copy-and-paste repeatedly). Therefore, it is recommended to create functions for every bit of code that is intended to be used more than once.

MATLAB toolboxes used in this chapter:
Image Processing Toolbox

Index of the in-built MATLAB functions used:

colormap	imagesc	numel	strcmp
dicominfo	isdicom	num2str	struct2cell
dicomread	linspace	plot	sum
dicomreadVolume	max	regionprops	writecell
disp	mean	return	xlabel
double	meshgrid	size	ylabel
fieldnames	min	sort	zeros
gca	NaN	squeeze	
hold	nanmean	std	

---

[1]If you do not have access to an ATCM phantom, use any other suitable phantom to start with.

## CHAPTER 12

Parsing and analyzing Radiation Dose Structured Reports

Robert Bujila

Medical Radiation Physics and Nuclear Medicine, Karolinska University Hospital, Stockholm, Sweden

CONTENTS

R ADIATION DOSE STRUCTURED REPORTS (RDSR) have become the preferred method for recording dose indices in diagnostic radiology. This chapter will show how the information in RDSR objects can be parsed, aggregated and analyzed to provide a meaningful overview of dose indices.

12.1 INTRODUCTION

Radiation dose in X-ray imaging is a major concern among healthcare professionals. To address concerns over the need to monitor/track dose indices in diagnostic radiology, the X-ray Radiation Dose Structured Report was introduced through Supplement 94 in the Digital Imaging and Communications in Medicine (DICOM) standard[98]. An RDSR object can be described as a machine readable collection of technique parameters and associated dose indices that were used during an examination. Unfortunately, the information in RDSR objects can be difficult for a "casual" user to navigate due to its structure.

The purpose of this chapter is to show how one could use MATLAB to parse the information contained in RDSR objects. After an initial simplificaton of RDSR objects, information of interest can be parsed and aggregated into a table. Statistical methods can then be applied to that data. Examples are given for Computed Tomography, however, the methods discussed in this chapter can be applied to other modalities that produce RDSR objects. Note, this chapter will not cover how RDSR objects are collected and assumes that the RDSR objects have been anonymized beforehand.

12.2 STRUCTURE OF RDSR OBJECTS

The structure of an RDSR object is defined in parts 3 and 16 of the DICOM standard[99, 100]. General information, for example, patient, study, etc., is available in a way similar to image objects. The technique parameters and dose indices on the other hand are grouped together in containers. These containers can be nested, meaning that a container can contain other containers which have a grouping of data points. A high level example of what an RDSR object used in CT can look like and where different pieces of data are located is presented in Figure 12.1. Note, in this example, only one CT Acquisition Container is presented, however, if multiple exposures are used, each exposure would get their own CT Acquisition Container.

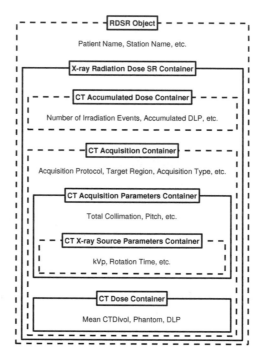

Figure 12.1: Visualization of the nested structure of CT RDSR objects and where different items of information are located.

The available information that is located prior to the X-ray Radiation Dose SR Container is univariate in the sense that one item of information is mapped to one attribute, for example, the patient's name is associated with the tag (0010, 0010). However, the technique parameter/dose index information contained within the different containers are codified using content sequences. With this implementation, the content sequences use several general attributes to convey one item of information. An example of the content sequence used for the mean $CTDI_{vol}$ is presented in Table 12.1. When looking at Table 12.1 one needs to at least consider the Code Meaning and Numeric Value attributes to put the mean

CTDI$_{vol}$ into context. Note, similar content sequences will be used for the different technique parameters/dose indices. The content sequences can be formatted and structured differently depending on the Value Type (see Table 12.1). Value types can be for example CONTAINER, CODE, TEXT, DATETIME and UIDREF (unique identifier).

Table 12.1: Example of a content sequence to convey the numerical value for the Mean CTDI$_{vol}$ of a single irradiation event. The angled bracket, >, denotes the level within the nested structure and VR is the Value Representation of the attribute.

Attribute Tag	Attribute Description	VR	Value
(0040,a730)	Content Sequence		
> (0040,a010)	Relationship Type	CS	Contains
> (0040,a040)	Value Type	CS	NUM
> (0040,a043)	Concept Name Code Sequence		
> > (0008,0100)	Code Value	SH	113830
> > (0008,0102)	Code Scheme Designator	SH	DCM
> > (0008,0104)	Code Meaning	LO	Mean CTDIvol
> (0040,a300)	Measured Value Sequence		
> > (0040,a30a)	Numeric Value	DS	8.88

12.2.1 MATLAB intepretation of RDSR objects

As with other DICOM objects, RDSR objects can be imported into the MATLAB workspace using the **dicominfo** command from the *Image Processing Toolbox*. However, due to the extensive use of containers and content sequences in the RDSR, the resulting *struct* (with its nested structure) can be difficult to extract information of interest from. Below is an example of how one can load an RDSR object and navigate to the tube voltage (kVp) that was used in the third irradiation event of an RDSR object. Note, the below method can be unreliable when scripting since the structure (locations of information of interest) can change depending on how many irradiation events and which techniques were used in an examination:

```
dcmInfo = dicominfo('myCTRDSR.dcm')
kvp = dcmInfo.ContentSequence. ... % RDSR container
        Item_15.ContentSequence. ... % CT Acquisition container for 3rd event
        Item_6.ContentSequence. ... % CT Acquisition Parameters container
        Item_8.ContentSequence. ... % CT Source Parameters container
        Item_2.MeasuredValueSequence.Item_1.NumericValue % kVp
```

12.2.2 Simplifying the contents of an RDSR object in MATLAB

To alleviate some of the difficulties when navigating RDSR objects, the structure can be simplified. One way of simplifying the structure of the RDSR data is to loop through each content sequence in the nested set of containers and extract the TEXT/CODE/DATETIME/NUM/UIDREF variables. In each iteration of the loop, the variables can be added to a struct, where the variable names are the Code Meaning of the content sequence (see Table 12.1) to understand the context. In order to traverse the RDSR containers in a systematic way, a recursive function can be used where the function can call itself when it encounters a container (in effect going to the next level in the nested structure). Since identical containers can exist in the same nested level, an ordinal numbering scheme is used to differentiate between the containers:

```matlab
function simpleRdsr = simplifyRdsr(ContentSequence)
% A function to simplify the structure of an RDSR object
% Input: Content Sequence from RDSR object (e.g., dcmInfo.ContentSequence) [struct]
% Output: Simplified RDSR struct [struct]

% Create an empty struct
simpleRdsr = struct;

% Get the fieldnames of the input ContentSequence
fieldNamesInSequence = fieldnames(ContentSequence);

% Loop over field names in ContentSequence
for fieldNameNumber = 1:numel(fieldNamesInSequence)

    % Get the concept name and value type
    fieldData = ContentSequence.(fieldNamesInSequence{fieldNameNumber});
    conceptName = fieldData.ConceptNameCodeSequence.Item_1.CodeMeaning;

    % Remove spaces and hyphens from conceptName
    conceptNameFriendly = regexprep(conceptName, {' ', '-'}, '');
    valueType = fieldData.ValueType;

    % Get the value depending on value type
    switch valueType
        case 'CONTAINER'
            % If valueType is CONTAINER use recursion to go to next level
            value = simplifyRdsr(fieldData.ContentSequence);
        case 'NUM'
            value = fieldData.MeasuredValueSequence.Item_1.NumericValue;
        case 'UIDREF'
            value = fieldData.UID;
        case 'TEXT'
            value = fieldData.TextValue;
        case 'DATETIME'
            value = fieldData.DateTime;
        case 'CODE'
            value = fieldData.ConceptCodeSequence.Item_1.CodeMeaning;
    end %switch

    % At the same level in the RDSR, there can be multiple sequences with
    % the same name. Here each sequence is numbered.
    currentFieldNames = fieldnames(simpleRdsr);

    % See how many fieldnames in the current level match the concept name
    % of the sequence that is being simplified.
    currentFieldNameMatch = startsWith(currentFieldNames,...
                                        conceptNameFriendly);
    % startsWith returns a logical array. By summing that array, we get the
    % number of matches.
    numMatches = sum(currentFieldNameMatch);
```

(code continues on next page)

```
% Number each concept name according to the number of occurences in the
% same level.
if numMatches == 0
    variableName = [conceptNameFriendly '_1'];
else
    variableName = [conceptNameFriendly '_' num2str(numMatches + 1)];
end % if


% Add the value to the simpleRdsr struct
% Note, because of the recursion that is used, the different levels in
% the RDSR objects will be maintained in the simpleRdsr struct.
simpleRdsr.(variableName) = value;
end % for loop over fieldNamesInSequence
end % function
```

Below is an example of how the **simplifyRdsr** function can be used to simplify an RDSR object and extract the tube voltage (kVp) from the third irradiation event. Note that **simplifyRdsr** is passed the top-level content sequence (the contents of the X-ray RDSR container) and returns a nested structure with its contents. It can be of interest to compare the clarity of this method compared to when not simplifying the structure of the RDSR (see the previous subsection). The use of **simplifyRdsr** reduces the likelihood of error as the nested structure and nomenclature is in logical agreement with the nesting of containers in Figure 12.1:

```
dcmInfo = dicominfo('myCTRDSR.dcm');
rdsrContainer = dcmInfo.ContentSequence;
simpleRdsr = simplifyRdsr(rdsrContainer);
kvp = simpleRdsr. ...
    CTAcquisition_3. ...
    CTAcquisitionParameters_1. ...
    CTXRaySourceParameters_1.KVP_1
```

12.3 PARSING RDSR OBJECTS

12.3.1 Parsing a single RDSR object

A function can be created to parse data (extract information of interest) from a single RDSR object that has been simplfied. Note, the structure and content of RDSR objects can vary greatly between vendors and especially modalities. It can be beneficial to consult the X-ray systems DICOM conformance statement to get a better understanding of the expected content. Once simplified, the simplified RDSR struct can be inspected in the Workspace of the MATLAB Desktop Environment in order to get an understanding of the items of information that are available in a particular RDSR object. In this example, the tube voltage (kVp), collimation, rotation time, pitch, $CTDI_{vol}$, CTDI Phantom, and DLP from all spiral events are of interest. The function loops over all irradiation events and if the irradiation event is a spiral scan, it puts the information of interest into a *table*. Additionally, the patient weight is also parsed and included in the table output:

```matlab
function parsedData = parseCtRdsrSpiral(fileName)
% Function to parse technique parameters and dose indices from irradiation
% events using spiral acquisition mode
% Input: fileName [string]
% Output: parsedData [table]

dcmInfo = dicominfo(fileName); % Read DICOM info
patientWeight = dcmInfo.PatientWeight; % Get patient weight
rdsrContainer = dcmInfo.ContentSequence; % Get RDSR Container
rdsrStruct = simplifyRdsr(rdsrContainer); % Simplify RDSR

% Get the number of irradiation events from accumulated dose data container
accDoseData = rdsrStruct.CTAccumulatedDoseData_1;
numEvents = accDoseData.TotalNumberofIrradiationEvents_1;

% Count spiral events in RDSR
cnt = 1;

% Loop over each irradiation event
for eventNum = 1:numEvents
    % Get the acquisition (scan) type
    eventData = rdsrStruct.(['CTAcquisition_' num2str(eventNum)]);
    scanType = eventData.CTAcquisitionType_1;

    % if scanType is spiral acquisition, parse data
    switch scanType
        case 'Spiral Acquisition'
            % Parse Acquisition Parameters Container
            acqParams = eventData.CTAcquisitionParameters_1;
            collimation(cnt) = acqParams.NominalTotalCollimationWidth_1;
            pitch(cnt) = acqParams.PitchFactor_1;

            % Parse Xray Source Parameters Container
            xraySrcParams = acqParams.CTXRaySourceParameters_1;
            kvp(cnt) = xraySrcParams.KVP_1;
            rotTime(cnt) = xraySrcParams.ExposureTimeperRotation_1;

            % Parse CTDose Container
            ctDose = eventData.CTDose_1;
            ctdi(cnt) = ctDose.MeanCTDIvol_1;
            phantom{cnt} = ctDose.CTDIwPhantomType_1;
            dlp(cnt) = ctDose.DLP_1;

            % Add patient weight
            weight(cnt) = patientWeight;

            cnt = cnt + 1;
    end % switch scanType
end % for exposureNum

% Add data to table
parsedData = table(kvp, collimation, rotTime, pitch, ctdi, phantom, dlp, weight);
end % function
```

12.3.2 Parsing multiple RDSR objects

A script can be created to loop over a collection of RDSR objects, for example, located in a directory. Initially, the contents of a folder is evaluated and a loop is initiated to iterate through each file in that folder. Each file is inspected and if it is DICOM, the `parseCTRdsrSpiral` function is run on that file. In each loop, the parsed data for each RDSR object is appended to a table. Note, the loop has been constructed in a way that is compliant with the *Parallel Computing Toolbox* (i.e., the parsed data is appended to the table in a non-deterministic order). By replacing **for** with **parfor** in the script below, the parsing can be done on multiple CPU cores.

```
% The directory where CT RDSR objets are located
rdsrDirectory = 'demoSR';

% Get contents of directory
dirContents = dir(rdsrDirectory);

% Create a data table to aggregate the parsed data
dataTable = [];

% Loop over directory contents
% This code is setup to work with the parallel processing toolbox, if you
% do have that toolbox, replace for with parfor.
for fileNumber = 1:numel(dirContents)

    % Create path/file name
    fileName = fullfile(rdsrDirectory, dirContents(fileNumber).name);

    % If "fileName" is not a directory and is DICOM proceed
    if ~dirContents(fileNumber).isdir && isdicom(fileName)
        parsedData = parseCtRdsrSpiral(fileName);

        % Append parsed data to dataTable
        dataTable = [dataTable; parsedData];
    end
end
```

12.4 ANALYZING PARSED RDSR DATA

In these examples, 306 RDSR objects for the "Routine CT Chest w/o IV contrast" protocol on one scanner have been collected during a 6 month period at the St. Elsewhere Hospital. The X-ray technologists have been instructed to input the patients' weight manually on the CT console at scan time so that this information will be included in the RDSR objects. Using a single CPU core, it took ~70 seconds to parse all of the RDSR objects on a laptop computer. Using the *Parallel Computing Toolbox*, it took ~50 seconds on that same computer with 2 CPU cores.

12.4.1 Diagnostic Reference Levels

One task that a medical physicist in diagnostic radiology may encounter is that of comparing local practice against Diagnostic Reference Levels (DRL)[101]. Using national guidelines given by, for example, the Swedish Radiation Safety Authority (SSM)[102], the calculated DRL quantity should be calculated as the mean $CTDI_{vol}$ and DLP for at least 20 patients that weigh between 60 and 90 kg. Since the patients' weight is available in our parsed data, the data points can be binned into categories of <60 kg 60-90 kg and >90 kg using MATLAB's `discretize` function.

```
% Categorize the patient's weight
dataTable.weightCategory = ... % Append categorization to datatable
                    discretize(dataTable.weight,... % Use function on weights
                    [0, 60, 90, inf],... % Bin edges
                    'categorical',... % Array type
                    {'<60 kg', '60-90 kg', '>90 kg'}); % Categories
```

Now that each of the data points has been binned into weight categories, summary statistics over different groupings of the data can be made using the `grpstats` function from the *Statistics and Machine Learning Toolbox*. In this example, the data is grouped with the available technique parameters as well as weight category. The summary statistics will include the mean patient weight, $CTDI_{vol}$ and DLP of each grouping.

```
statArray = grpstats(dataTable,...
            {'kvp', 'collimation', 'rotTime',...
            'pitch', 'phantom', 'weightCategory'},... % Group by these variables
            'mean',... % Summary statistics of variables of interest
            'DataVars', {'weight', 'ctdi', 'dlp'}) % Variables of interest
```

The function `grpstats` generates a table where each row represents a single grouping (by group variable value and category). Table 12.2 shows the results from the grpstats table where kVp, collimation, rotation time and phantom have been omitted since they were invariant. The summary statistics (mean value) is located to the right in this table. The DRL quantity is the second row from the bottom. Inspecting the resulting table shows that, in general, the same technique parameters were used for all "CT Chest w/o IV Contrast" scans. However, the spiral pitch factor was varied for four patients that were binned in the >90 kg category. Note, while only the mean was used for analysis in the `grpstats` function, additional statistics can be included (e.g., median, standard deviation, 75th percentile, etc.).

Table 12.2: Results from the *grpstats* function on the parsed RDSR data.

Pitch	Weight Category (Group Count)	Mean Weight [kg]	Mean $CTDI_{vol,32cm}$ [mGy]	Mean DLP [mGycm]
0.8	>90 kg (1)	127.0	22.7	803.0
0.9	>90 kg (1)	125.0	18.1	646.1
1.05	>90 kg (1)	111.0	14.6	573.6
1.1	>90 kg (1)	96.0	10.4	403.2
1.2	<60 kg (72)	53.9	4.9	176.2
1.2	60-90 kg (180)	73.6	7.4	268.7
1.2	>90 kg (50)	98.8	10.8	401.9

12.4.2 Evaluating Automatic Tube Current Modulation (ATCM)

ATCM is an important feature on CT scanners. It modulates the tube current at different positions during a scan to provide a target image quality throughout the scan[97]. The data that has been parsed in this chapter can be used to evaluate the performance of the scanner's ATCM in terms of how the $CTDI_{vol}$ varies with patient weight.

The table array data type that has been used throughout this chapter makes it easy to slice data in different ways. The script below can be used to generate a plot that shows

CTDI$_{vol}$ vs patient weight for: (1) all data points that have a pitch of 1.2 and (2) all data points that do not have a pitch of 1.2 (see Table 12.2).

```
plot(dataTable(dataTable.pitch == 1.2,:).weight,... % weights with pitch 1.2
     dataTable(dataTable.pitch == 1.2,:).ctdi,... % ctdi with pitch 1.2
     'k.',...
     'DisplayName', 'Pitch == 1.2')
hold on
plot(dataTable(dataTable.pitch ~= 1.2,:).weight,... % weights with pitch other than
     1.2
     dataTable(dataTable.pitch ~= 1.2,:).ctdi,... % ctdi with pitch other than 1.2
     'ko',...
     'DisplayName', 'Pitch != 1.2')
legend('location', 'NorthWest') % Uses the names given with DisplayName
xlabel('Patient Weight [kg]')
ylabel('CTDI_{vol,32cm} [mGy]')

% Use cell array if you want to have a title on multiple rows
title({'CT Room 1', 'CT Chest w/o IV Contrast'})
```

Figure 12.2 demonstrates that the ATCM settings that have been selected on the scanner allows the tube current to modulate over a wide range for a large range of patient sizes. Figure 12.2 also shows that the pitch was altered for larger patients.

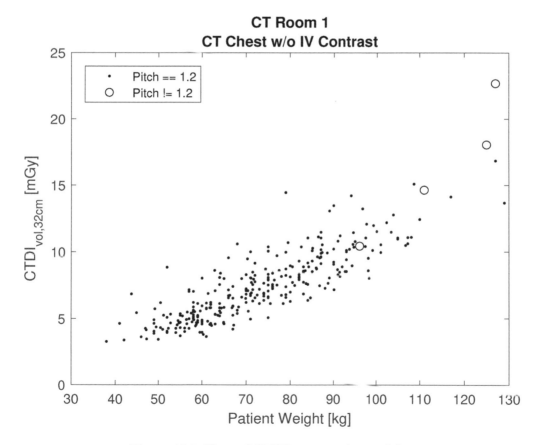

Figure 12.2: Plots of CTDI$_{vol}$ vs. patient weight.

12.5 CONCLUSIONS

RDSR objects are an important feature on X-ray systems that report the technique parameters and dose indices of X-ray examinations. When parsed, the data in RDSR objects can be aggregated to track/monitor X-ray examinations from a radiation safety perspective. This chapter provides a method to first restructure the RDSR objects into a form that is easier to manage than when using MATLAB's native functionality. After RDSR objects are restructured, they can then be parsed to extract relevant data. Two examples were then given on how the parsed data could be used to: (1) calculate DRL quantities and (2) evaluate ATCM.

While this chapter has mainly focused on the CT modality, the restructuring of the RDSR objects will likely work on modalities other than CT (e.g., X-ray Angiography or planar radiography). The script that was used to parse the RDSR *structs* can be adapted to these other modalities after the user has become familiar with the contents in those RDSR objects.

MATLAB toolboxes used in this chapter:
Image Processing Toolbox
Statistics and Machine Learning Toolbox

Index of the in-built MATLAB functions used:

dicominfo	grpstats	num2str	table
dir	hold	plot	title
discretize	isdicom	regexprep	xlabel
fieldnames	legend	struct	ylabel
fullfile	numel	sum	

## CHAPTER 13

Method of determining patient size surrogates using CT images

Christiane Sarah Burton

Department of Radiology, Boston Children's Hospital and Harvard Medical School, Boston, MA, 02215, USA

CONTENTS

T HE PURPOSE of this chapter is to show how a medical physicist can calculate the size-specific dose estimate (SSDE) using patient size surrogates from the DICOM images and information about the scan from DICOM metadata. The example will show how DICOM images can be read in and used to determine the water-equivalent diameter (WED) and anterior-posterior (AP) and lateral (LAT) dimensions.

13.1 INTRODUCTION

Radiation exposure from CT has become an increasing concern within the medical imaging community. For medical physicists, it has always been a challenge to balance the ALARA (As Low As Reasonably Achievable) principles with achieving the best possible image quality. Tracking patient dose is important for medical physicists who work in radiology. Therefore having a good quantifiable estimate of the absorbed dose to the patient is needed. Unlike $CTDI_{vol}$ and dose length product (DLP), the size-specific dose estimate (SSDE) gives an estimate of the patient's absorbed dose. The $CTDI_{vol}$ is a measure of the system's output, specifically for 32 and 16 cm diameter cylidrical phantoms made of poly(methyl methacrylate) (PMMA) in a contiguous axial and helical examination[103]. For a medical physicist, it is more meaningful to track and report the SSDE as a measure of patient dose.

The SSDE is the product of the normalized dose coefficient (NDC) and the $CTDI_{vol}$[9]. The NDC is calculated using patient size surrogates such as: lateral width (LAT), anterior-posterior width (AP), effective diameter and water-equivalent diameter (WED)[10]. It is preferable to use the WED because it takes into account patient attenuation and is therefore more appropriate for scaling dose, particularly for the thorax region of a chest scan[104].

All of these patient size surrogates can be calculated from CT axial images or CT localizer radiographs. The LAT and AP widths may be extracted directly from the CT axial image. For CT localizer radiographs, in order to get an accurate estimate of LAT and AP dimensions, a magnification correction must be applied[105]. The effective diameter is calculated by taking the geometric or arithmetic means of the LAT and AP widths[9].

From an axial CT image or localizer radiograph, the WED is calculated from the water-equivalent area. This area explicitly accounts for the fact that the the linear attenuation coefficient inside a patient (and hence CT-numbers in Hounsfield Units) differ from that of water. Methods for both cases (axial and localizer) are outlined in the Report of AAPM Task Group 220[10]. Note that unlike CT numbers, the pixel values of a CT localizer radiograph are not reported in absolute units and the relationship between attenuation and pixel value can vary among scanner models and manufacturers. A calibration of CT localizer radiograph pixel values in terms of water attenuation is therefore required. If the WED is needed from the CT localizer radiograph then it is recommended that the estimate be implemented only by those who have carefully determined the necessary calibration.

This chapter will demonstrate how patient size surrogates can be extracted from patient DICOM images. The demonstration is provided for CT axial images. A table removal method will be shown, as inclusion of the table in WED estimates is not recommended. Another important parameter is the ellipticity ratio (ratio of LAT/AP widths), which can vary throughout the patient scan, and can even be the different for two CT slices with the same WED[106]. The CT vendor's automatic exposure control (AEC) takes into account the path that the x-ray travels, meaning that the technique factors the AEC selects is based on the estimated thickness of the patient. Understanding the behaviour of the AEC therefore involves knowing the ellipticity ratio for various body regions. This chapter will therefore also demonstrate the simple calculation of ellipticity ratio.

Note that while the axial approach is generally preferable when possible, an advantage of the localizer method is that it can be applied *prior* to the acquisition of axial data (or when axial images are not available). While methods for extracting the relevant size metrics from localizers are not discussed further, the reader can fine more details on that approach in the American College of Radiology Dose Index Registry's (ACR DIR) method, which includes table removal[107].

13.2 STRUCTURE OF THE CODE

In the following sections, MATLAB code will be presented which performs all of calculation tasks automatically. For simplicity, input from the user is needed to locate the image directory where the patient data is stored. To use this feature, each patient folder must contain CT axial images for a single scan. Using this approach, many scans can be extracted and processed and all information placed into a single data structure (called `exam`).

The section of code below requests a data location from the user (using `uigetdir`). Using this input, a nested loop is conducted through all folders (examinations) and files (slices). Each DICOM image and its metadata are read in and stored in `dcmImage` and `dcmInfo`, respectively. The size metrics for axial images are calculated through the script `calcSizeMetrics`. The SSDE estimates for the examinations are calculated by the script `calcSSDE`. Note that since `calcSizeMetrics` and `calcSSDE` are scripts rather than functions, they share variable scope with the main script (they have a common *workspace* in MATLAB terminology).

```matlab
% Locate the directory where the patient folders are stored
defaultDirectory = 'C:\PatientDataSet';
dataLocation = uigetdir(defaultDirectory);

% Whether to remove table: 'y' (yes) or 'n' (no)
tableRemoval = 'n';

% Exit if user aborted the uigetdir dialog
if dataLocation == 0, return, end

% Store the path to each folder and start exam count
folderNames = dir(dataLocation);
iExam = 1;

% Loop over all folders in dataLocation
for iFolder = 1:(size(folderNames,1))
  % Ignore top-level (.) and parent (..) folders for dataLocation
  if strcmp(folderNames(iFolder).name, '.') || ...
    strcmp(folderNames(iFolder).name, '..')
    continue
  end

  % Store the path to each file in the folder
  fileNames = dir(fullfile(dataLocation, folderNames(iFolder).name));

  iSlice = 1; % Start slice count
  for iFile = 1:(size(fileNames,1)) % Loop over all files in a folder
    % Ignore top-level (.) and parent (..) folders of fileNames
    if strcmp(fileNames(iFile).name, '.') || ...
      strcmp(fileNames(iFile).name, '..')
      continue
    end

    % Full path to DICOM file
    file = fullfile(dataLocation,folderNames(iFolder).name,...
    fileNames(iFile).name);
    % Read DICOM meta-data
    dcmInfo = dicominfo(file);
    % Read DICOM image and convert to CT-numbers
    dcmImage = dcmInfo.RescaleSlope.*double(dicomread(file)) ...
      + dcmInfo.RescaleIntercept;

    % Only analyze the file if it is an axial slice
    if (count(dcmInfo.ImageType,'ORIGINAL\PRIMARY\AXIAL') > 0)
      calcSizeMetrics; % Execute MATLAB script, calcSizeMetrics.m
      iSlice = iSlice + 1; % Increment slice count
    end
  end

  iExam = iExam + 1; % Increment folder count
end

clearvars -except exam % Clean up MATLAB workspace

calcSSDE; % Execute MATLAB script, calcSSDE.m
```

13.3   CALCULATING SIZE METRICS FROM CT AXIAL IMAGES

13.3.1   The size metrics

This section describes the calculations carried out by the `calcSizeMetrics` script. A method is presented for extracting patient size surrogates (AP and LAT widths) from CT axial scans and the calculation of effective diameter and the ellipticity ratio. A method for extracting the mean Hounsfield Unit from a region of interest (ROI) in a CT scan and a water-equivalent diameter (WED) calculation is shown. A method for table removal is also described. The reader is provided with example code here along with a detailed explanation of the methodology in the following subsections. All the pertinent data for each slice, including the entire DICOM metadata, is stored in a structure called `exam`.

13.3.2   Calculating AP and LAT dimensions

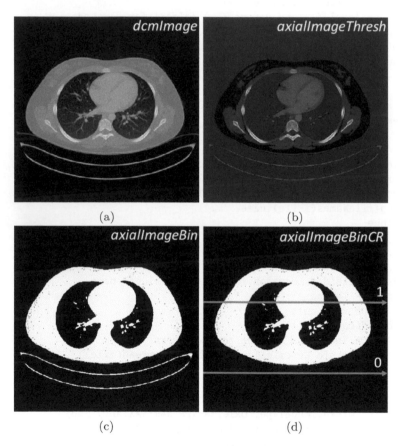

Figure 13.1:   (a) The original CT axial image, (b) CT axial image with a threshold of ≤ -150 HU set to zero applied, (c) binary CT axial image with the remaining values over -150 set 1, (d) binary CT axial image with table removed. In the last sub-figure, two rays are shown from those used to determine the lateral width. One ray depicts a path through the patient (value 1) and the other through air only (value 0).

Figure 13.1 provides a visual for how `calcSizeMetrics` extracts AP and LAT dimensions from a single CT axial image. Figure 13.1a shows the original CT axial image. Figure 13.1b shows the image after a threshold of -150 is applied, to exclude the air in the background, leaving the shape of the patient. The image is binarized by assigning a value of 1 to any value that is not zero, resulting in the image shown in Figure 13.1c). However, the table must be removed from the image so that it is not included in the estimation of the patient size surrogates LAT and AP. Using MATLAB's in-built bwareaopen function, all connected components or objects in the image that contain fewer than a specified number of pixels (in this case, 5000 pixels) are removed from the binary image, to eliminate any objects outside of the contours of the patient (e.g., the table, EKG leads, tubing and cloths/blankets). The result is shown in Figure 13.1d.

A summation along each row or column in the image is then conducted, where any values over zero are assigned a 1 if the patient is present along the line or a zero if the patient is not present along the rows and columns (see Figure 13.1d). These binary values are then stored in arrays and each array is summed to calculate the LAT or AP widths in pixels. The sums of these arrays are then multiplied by the pixel spacings values from DICOM tag (0028,0030) in the structure dcmInfo, to obtain the widths in mm (widthLAT and widthAP). Figure 13.2 illustrates LAT and AP widths calculated for the ATCM phantom discussed in Chapter 11 (*Tools for CT physicists: automating quality controls and evaluating automatic tube current modulation*), using the code of this chapter. The true dimensions of the sections of the phantom are known (300 mm × 200 mm, 350 mm × 233 mm and 250 mm × 167 mm) and in close agreement with the reconstructed values.

The calculation of effective diameter is performed in two ways. The first (diaEff1) takes the square-root of the product of the LAT and AP widths and the second (diaEff2) takes the summation of LAT and AP widths divided by 2. The ellipticity ratio (ellipRat) is calculated as the ratio of LAT (numerator) to AP (denominator).

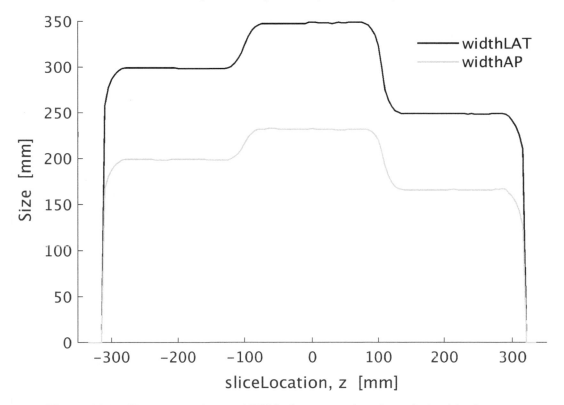

Figure 13.2: Size metrics for an ATCM phantom using the code in this chapter.

```matlab
% Set any padded values to CT-number of air
dcmImage(dcmImage <= -1024) = -1000;
% Make a binary image of the patient
axialImageThresh = dcmImage;
axialImageThresh(axialImageThresh <= -150) = 0; % Exclude low densities
axialImageBin(axialImageThresh ~= 0) = 1; % Include the rest

% Exclude connected components smaller than 5000 pixels (e.g., table)
axialImageBinCR = bwareaopen(axialImageThresh,5000);

% Binary projection along columns and width calculated from non-zero
% elements of projection
projAP = sum(axialImageBinCR,2) > 0;
widthAP = sum(projAP)*dcmInfo.PixelSpacing(1);
% Binary projection along rows and width calculated from non-zero elements
% of projection
projLAT = sum(axialImageBinCR,1) > 0;
widthLAT = sum(projLAT)*dcmInfo.PixelSpacing(2);

% Effective diameter is calculated using two formulas given in
% AAPM Report 204
diaEff1 = sqrt(widthLAT*widthAP); % Geometric mean
diaEff2 = (widthLAT + widthAP)/2; % Arithmetic mean
ellipRat = widthLAT/widthAP; % Ellipticity ratio

% Make a new binary image of patient including low densities inside patient
if tableRemoval == 'y' % Option to remove background and table from image
  % Fill in the low density regions inside patient, e.g lungs
  binImage = imfill(axialImageBinCR,'holes');
  areaROI = sum(sum(binImage))*dcmInfo.PixelSpacing(1)*...
  dcmInfo.PixelSpacing(2);
elseif tableRemoval == 'n' % Option to keep background and table
  % Loop through pixels in image
  s = size(dcmImage);
  binImage = zeros(s,'logical');
  for iRow = 1:s(1)
    for iCol = 1:s(2)
      if ((iRow-s(1)/2)^2+(iCol-s(2)/2)^2 < (dcmInfo.Width/2)^2)
        binImage(iRow,iCol) = 1; % Assign a non-zero value if inside FOV
      end
    end
  end
  % Area of the FOV
  areaROI = pi*(double(dcmInfo.Width)/2)^2*...
  double(dcmInfo.PixelSpacing(1))*double(dcmInfo.PixelSpacing(2));
end

% Mean CT number in patient
meanCTNum = mean2(dcmImage(binImage));
```

(code continues on next page)

```
% Water equivalent area and diameter calculation formula from AAPM Report 220
areaWaterEq = 0.001 * meanCTNum * areaROI + areaROI;
diaWaterEq = 2 * (areaWaterEq/pi)^(1/2);

% Store data in a structure called exam
exam(iExam).sliceLocation(1,iSlice) = dcmInfo.SliceLocation;
exam(iExam).diaWaterEq(1,iSlice) = diaWaterEq;
exam(iExam).ellipRat(1,iSlice) = ellipRat;
exam(iExam).diaEff(1,iSlice) = diaEff1;
exam(iExam).widthLAT(1,iSlice) = widthLAT;
exam(iExam).widthAP(1,iSlice) = widthAP;
exam(iExam).ctdiVol(1,iSlice) = dcmInfo.CTDIvol;
% Store entire DICOM header as a cell
exam(iExam).dicomHeaderAxial(1,iSlice) = {dcmInfo};
```

13.3.3 Table considerations and calculating Water Equivalent Diameter

The AAPM Report 220 explicitly states that the table should be removed prior to calculating WED, as this is the preferred method for SSDE calculations. In all WED publications that we are aware of to date, the table has not been removed. While the added affect of the table may not contribute substantially to the overall WED[106, 108], to get the best possible NDC and SSDE estimates with most relevance to the patient, the table should be removed.

If the variable removeTable has been set to 'y', then, the binary image generated by the bwareaopen function is used. The table has been eliminated from the images (as mentioned in Section 13.3.2) so that it will not be factored into the AP and LAT extraction. Using the output image from bwareaopen, the imfill function with 'holes' option is then used to fill in any low density cavities (e.g., the lungs). This binary image (binImage) can then be used as a mask representing all pixels inside the patient. If the variable removeTable has been set to 'n', the whole reconstruction field-of-view (assumed to be centered and of diameter dcmInfo.Width) is used to create the binary image. The MATLAB code extracts the Hounsfield Units (HU) from the defined circular region of interest. This circular ROI is created using a nested *for* loop. Although nested loops are not time efficient in execution, this approach is simple to code and understand. AAPM Report 220 provides the formula for the WED, which uses both the mean CT number and area of the ROI. Note that in some cases, when the reconstructed image center is not at isocenter, this ROI could contain "padding" values, for example, -3024 HU. This issue was handled by the first line of code in the calcSizeMetrics, where all voxels with a value of ≤ -1024 HU were reset to values of -1000 HU to simulate air. The use of padding values is common to most CT vendors, but the padding value may differ. Failure to correct for this before doing a WED calculation could significantly decrease the WED values.

13.4 CALCULATING THE SIZE-SPECIFIC DOSE ESIMATE

To use the information stored in the structure exam, the script calcSSDE is executed. Any data extracted from the structure can be plotted or printed in MATLAB or be stored in an Excel spread sheet. The CTDI$_{vol}$ can be extracted and used with the NDC to calculate SSDE. In the example code below, the average NDC in the central 20% of each examination is estimated, using both the effective-diameter (diaEff) and WED (diaWaterEq). The average CTDI$_{vol}$ in the same region is also calculated and used to convert NDC to SSDE.

Before these calculations, the data stored in **exam** for every examination is reordered so that the slice location is ordered with increasing value (using MATLAB's **sort** function).

```
% Reorder the slices in each exam so we go from low to high z-position
for iExam = 1:length(exam)
  [~,b] = sort(exam(iExam).sliceLocation); % Array b holds the reordering
  % Iterate through the fieldnames of the structure
  for fieldName = fieldnames(exam(iExam))
    fieldValues = exam(iExam).(fieldName{1});
    % Reorder only if the field has a value for each slice
    if numel(fieldValues) == numel(b)
      exam(iExam).(fieldName{1}) = fieldValues(b);
    end
  end
end


% For each exam, average the size metric and CTDIvol over middle 20% of
% scan
for iExam = 1:length(exam)
    numberSlices = numel(exam.sliceLocation);
    index1 = round(numberSlices/2 - 0.2*numberSlices/2);
    index2 = round(numberSlices/2 + 0.2*numberSlices/2);
    midDiaEff(iExam) = mean(exam(iExam).diaEff(index1:index2));
    midDiaWaterEq(iExam) = mean(exam(iExam).diaWaterEq(index1:index2));
    midCtdiVol(iExam) = mean(exam(iExam).ctdiVol(index1:index2));
end

% Fit coefficients from AAPM Report 204 (for CTDIvol 32cm)
a = 3.704369;
b = 0.03671937*0.1;

% Estimate the SSDE using effective diameter
ssdeDiaEff = a*exp(-b*midDiaEff).*midCtdiVol; % mGy
% Estimate the SSDE using water equivalent diameter
ssdeDiaWaterEq = a*exp(-b*midDiaWaterEq).*midCtdiVol; % mGy
```

13.5 CONCLUSION

Vendors will now be required to display SSDE and WED either before or after the CT scan, however their method of obtaining this data may not be revealed. This chapter has provided the reader with a detailed description of code that can be used to calculate patient size surrogates and the SSDE for an axial image, or to get an estimate of SSDE averaged over a specific region of a scan.

MATLAB toolboxes used in this chapter:
Image Processing Toolbox

Index of the in-built MATLAB functions used:

bwareaopen	dir	mean	sort
clearvars	double	mean2	sqrt
count	exp	numel	strcmp
dicominfo	fullfile	round	sum
dicomread	imfill	size	uigetdir

Reconstructing the exposure geometry in x-ray angiography and interventional radiology

Artur Omar

Medical Radiation Physics and Nuclear Medicine, Karolinska University Hospital, Stockholm, Sweden

Department of Oncology-Pathology, Karolinska Institutet, Stockholm, Sweden

CONTENTS

MEDICAL PHYSICISTS may find themselves entrusted with the daunting task of estimating the absorbed dose to a patient that has undergone a complex image-guided intervention consisting of hundreds of separate and different x-ray exposures. In this chapter, a framework for approaching such a task is outlined.

14.1 INTRODUCTION

In angiography and interventional radiology, the already challenging task of accurate patient dose estimation is compounded by the difficulty in obtaining detailed information about the exam-specific irradiation geometry and x-ray beam settings used. Nonetheless, considering the relatively high exposure associated with some complex interventions, the absorbed organ doses should be estimated with an accuracy that is adequate for assessing the risk of radiation induced tissue reactions, such as skin erythema and hair loss [109]. The International Atomic Energy Agency (IAEA) has in the international code of practice

(a) (b)

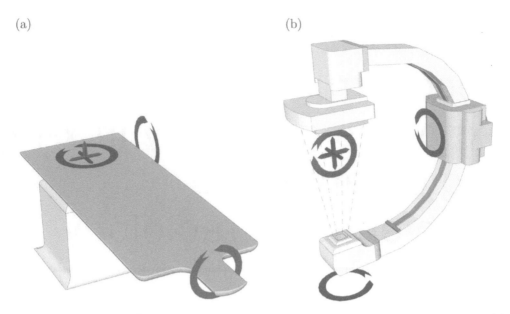

Figure 14.1: Schematic illustration of an x-ray angiography system that consists of (a) a patient table top, and (b) an x-ray tube and imaging detector mounted on a c-arm. Possible table top movement and tilting axes are shown, along with possible c-arm movement and rotation axes about the isocenter.

TRS-457[110] recommended an accuracy of 7% (95% confidence limits) for dosimetry measurements in diagnostic radiology when tissue reactions are expected. Although the quoted requirement relates to the accuracy of a directly measured dosimetry quantity, it indicates a general level of accuracy that should be pursued in a patient dose estimation. This level of accuracy is, however, difficult to achieve, especially if relying on simplified approximations for the exam-specific settings used (x-ray beam quality, irradiation geometry, etc.).

In order to allow for systematic and non-proprietary access to exam-specific exposure settings, the U.S. National Electrical Manufacturers Association (NEMA) has introduced diagnostic x-ray radiation dose structured reporting (RDSR)[111]; RDSR is, in accordance with IEC 60601-2-43[112], required to be supported on newly manufactured x-ray units. Although RDSR typically contains extensive information about the different exposure settings that have been used, including the patient table and x-ray beam geometry (see Figure 14.1), it tends to lack information about the position of the patient on the table top. Without such information, the geometrical alignment of the x-ray beam with the patient anatomy (i.e., the patient-beam alignment) cannot be accurately reconstructed, which may add a substantial uncertainty to a radiation dose calculation.

In this chapter, a method proposed in Ref. [113] for the reconstruction of the exposure geometry using only information contained in RDSR is demonstrated in a MATLAB implementation. The chapter focuses on two key aspects of the exposure geometry: (i) inferring the position of the target region (body part) in relation to the x-ray tube source (Section 14.3), and (ii) how to calculate the source-to-skin distance (Section 14.4). In a final section (Section 14.5), the reconstructed exposure geometry is used to determine the incident air kerma, a quantity that is often used as input to patient dose calculations. Since this requires an understanding of elementary vector algebra, lets first focus on how translation, scaling, and rotation of position vectors can be implemented in MATLAB.

14.2 ELEMENTARY VECTOR ALGEBRA

14.2.1 Translation

A translation (or displacement) $\mathcal{T}(t_x, t_y, t_z)$ can be applied to the vector $\mathbf{r} = (x, y, z)$ using vector addition,

$$\mathbf{r}' = \mathcal{T}(\mathbf{t})\mathbf{r} = \mathbf{r} + \mathbf{t} = (x + t_x, y + t_y, z + t_z) = (x', y', z'). \qquad (14.1)$$

This can be implemented as:

```
%% Vector translation
r = [1,1,1]; t = [1,2,3]; % An arbitrary position vector and translation.

rp = r+t;    % r'= [2,3,4].
```

14.2.2 Scaling

A scaling transformation $\mathcal{S}(s_x, s_y, s_z)$ can be applied to the vector $\mathbf{r} = (x, y, z)$ using element-wise multiplication (i.e., the Hadamard product),

$$\mathbf{r}' = \mathcal{S}(\mathbf{s})\mathbf{r} = \mathbf{s}.*\mathbf{r} = (s_x x, s_y y, s_z z) = (x', y', z'). \qquad (14.2)$$

This can be implemented as:

```
%% Vector scaling
r = [2,2,2]; s = [1,2,3]; % An arbitrary position and scaling vector.

rp = s.*r;   % r'= [2,4,6].
```

14.2.3 Rotation

A vector rotation in three-dimensional space can be described as a sequence of three basic rotations about the coordinate axes: an angle α around the x-axis, an angle β around the y-axis, and an angle γ around the z-axis. A positive rotation about one of the axes would be viewed as a counterclockwise rotation by an observer looking along the rotation axis towards the origin, that is, the right-hand rule for the positive sign of the angles. A rotation of position vector $\mathbf{r} = (x, y, z)$ can thus be expressed as,

$$\mathbf{r}' = \mathcal{R}(\alpha, \beta, \gamma)\mathbf{r} = \mathcal{R}_z(\gamma)\mathcal{R}_y(\beta)\mathcal{R}_x(\alpha)\mathbf{r} = (x', y', z'), \qquad (14.3)$$

where $\mathcal{R}(\alpha, \beta, \gamma)$ is the rotation matrix. This can be implemented in MATLAB using anonymous functions (i.e., functions that are not stored in a program file) in terms of the three basic rotations:

```
%% Anonymous functions for the basic rotations Rx(), Ry(), Rz() (rad):
Rx = @(alph) [ 1        , 0          , 0          ; ...
               0        , cos(alph), -sin(alph) ; ...
               0        , sin(alph),  cos(alph)   ];
Ry = @(bet) [ cos(bet),  0          , sin(bet) ; ...
              0        ,  1          , 0          ; ...
             -sin(bet),  0          , cos(bet)   ];
Rz = @(gam) [ cos(gam), -sin(gam) , 0          ; ...
              sin(gam),  cos(gam) , 0          ; ...
              0        ,  0          , 1          ];
```

$$\mathbf{r}' = R\left(\frac{\pi}{2},0,0\right)\mathbf{r} \qquad \mathbf{r}' = R\left(\frac{\pi}{2},\frac{\pi}{2},0\right)\mathbf{r} \qquad \mathbf{r}' = R\left(\frac{\pi}{2},\frac{\pi}{2},\frac{\pi}{4}\right)\mathbf{r}$$

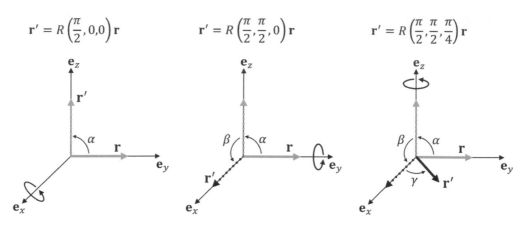

Figure 14.2: Rotation $\mathcal{R}(\alpha,\beta,\gamma)$ of a position vector $\mathbf{r} = (0,0.6,0)$ in the standard basis $\{\mathbf{e}_x, \mathbf{e}_y, \mathbf{e}_z\}$.

It should be noted that the implementation of a three-dimensional rotation is non-commutative (i.e., the order of the rotations matters). For instance, the rotation illustrated in Figure 14.2 is constructed in the following order:

```
%% Vector rotation
r = [0;0.6;0];                    % Position vector in fig. 14.2.
alph = pi/2; bet = pi/2; gam = pi/4; % Rotation angles in fig. 14.2.

rp = Rz(gam)*Ry(bet)*Rx(alph)*r; % r'= 0.6.*[cos(pi/4); sin(pi/4); 0].
```

Since $\mathcal{R}(\alpha,\beta,\gamma)$ is orthogonal, the inverse rotation is given by the transpose of the rotation matrix, meaning that a rotation can be reversed as,

$$\mathbf{r} = \mathcal{R}^{\mathrm{T}}(\alpha,\beta,\gamma)\mathbf{r}' = \mathcal{R}_x(-\alpha)\mathcal{R}_y(-\beta)\mathcal{R}_z(-\gamma)\mathbf{r}' = (x,y,z). \qquad (14.4)$$

14.3   RECONSTRUCTING THE PATIENT-BEAM ALIGNMENT

The absorbed dose to a patient can be estimated by reconstructing the exposure in a Monte Carlo simulation that includes a computational phantom model for the simulation of the patient anatomy. The exposure geometry can be reconstructed by matching the orientation and position of the phantom model to the patient's position and orientation on the table top. A method for this was suggested in Ref. [113] (referred to as the target-centric approach). The approach uses DICOM RDSR data to infer the patient position relative to the *x-ray system coordinate system* by identifying the primarily imaged body region, that is, the patient's target organ (e.g., the heart for cardiovascular interventions). The target organ can be located by taking advantage of the fact that although different exposure geometries are used as part of an x-ray procedure, most of the irradiation time is dedicated to visualizing the target region.

Hence, in order to locate the target organ, the location of the (x-ray beam/system) isocenter relative to the end of the patient table is formulated as a function of (irradiation) time,

$$\mathbf{r}_{\mathrm{iso}}(t) = \mathbf{r}_{\mathbf{b}}(t) - \mathbf{r}_{\mathbf{t}}(t), \qquad (14.5)$$

where $\mathbf{r}_{\mathbf{b}}$ and $\mathbf{r}_{\mathbf{t}}$ are, respectively, the beam isocenter position and the table head-end position in the x-ray system coordinate system. Then, given that the majority of the x-ray exposures aim to visualize the target organ and adjacent anatomy, the position of the target organ $[\mathbf{r_{target}} = (x_{\text{target}}, y_{\text{target}}, z_{\text{target}})]$ can be approximated as the median position of the isocenter $[\mathbf{r_{iso}} = (x_{\text{iso}}, y_{\text{iso}}, z_{\text{iso}})]$, as,

$$
\begin{aligned}
x_{\text{target}} &= \text{median}(\{x_{\text{iso}}(t) \mid t \in \mathcal{X}\}), \\
y_{\text{target}} &= \text{median}(\{y_{\text{iso}}(t) \mid t \in \mathcal{X}\}), \\
z_{\text{target}} &= \text{median}(\{z_{\text{iso}}(t) \mid t \in \mathcal{X}\}),
\end{aligned}
\tag{14.6}
$$

where the set $\mathcal{X}$ excludes time periods (i.e., exposure series) when the exposure is presumably not aimed at the visualization of the target region. This may for instance be the exclusion of time periods when the isocenter is located outside of the limitations imposed by a head fixation apparatus used for neurovascular procedures.

In order to implement the target-centric approach in MATLAB, relevant information about the exposure geometry must first be extracted from an RDSR file (this can be done using the techniques presented in *Chapter 12: Parsing and analyzing Radiation Dose Structured Reports*). Having extracted the relevant data, the target-centric approach can be implemented as:

```
%% Locating the target organ position
% [INPUT]:  RDSR     - data extracted from RDSR, rearranged into a struct
%                      (example input data is included in the .m file).
% [OUTPUT]: r_target - the position [x;y;z] of the patient's target organ
%                      in the x-ray system coordinate system.

% Spatial extent of head fixation apparatus relative to the table end [cm]:
headFixLimit_maxWidth  = 25; headFixLimit_minWidth  = -25;
headFixLimit_maxLength =  0; headFixLimit_minLength = -30;

% The beam isocenter relative to the end of patient table is determined:
r_iso = RDSR.xrayBeamPos-RDSR.tableEndPos; % According to eq. 14.5.

% The irradiation events with the beam isocenter located within the
% limits imposed by a head fixation apparatus are identified
% (relevant only for neurovascular interventions):
isEvent = r_iso(1,:) < headFixLimit_maxWidth  & ...
          r_iso(1,:) > headFixLimit_minWidth  & ...
          r_iso(3,:) < headFixLimit_maxLength & ...
          r_iso(3,:) > headFixLimit_minLength     ;
% And the associated (x,y,z) positions are defined:
[X,Y,Z] = deal( r_iso(1,isEvent), ... % deal is a native function to
                r_iso(2,isEvent), ... % distribute input variables
                r_iso(3,isEvent) );   % to outputs

% Each irradiation event is given an integer exposure duration [s]:
w = ceil(RDSR.ExposureDuration(isEvent)); % ceil rounds up to whole sec.

% The (x,y,z) positions are converted into a time series, i.e.,
% the positions are indexed in time order (1 s intervals):
[XinTime,YinTime,ZinTime] = deal([]);
```

(code continues on next page)

```
for i = 1:numel(w) % For each irradiation event
    % (x,y,z) of irradiation event "i" is converted into a time series:
    iX = repmat( X(i), 1, w(i) ); % repmat (a native function) returns
    iY = repmat( Y(i), 1, w(i) ); % an array of w(i) copies of e.g. X(i).
    iZ = repmat( Z(i), 1, w(i) );
    % The different time series (iX,iY,iZ) are iteratively concatenated:
    XinTime = cat( 2, XinTime, iX ); % cat (a native function)
    YinTime = cat( 2, YinTime, iY ); % concatenates e.g. XinTime and iX.
    ZinTime = cat( 2, ZinTime, iZ );
end

% The target organ position is approximated according to eq. 14.6:
r_target = [median(XinTime); ...
            median(YinTime); ...
            median(ZinTime)];
```

Having located the patient's target organ in the x-ray system coordinate system, it is now possible to reconstruct the exposure geometry for a computational phantom. The exposure at irradiation time t in the *computational coordinate system* (indicated with the superscript 'comp') can be reconstructed by translation of the coordinate system origin,

$$\mathbf{r}_{\mathrm{iso}}^{\mathrm{comp}}(t) = \mathbf{r}_{\mathrm{target}}^{\mathrm{comp}} + \left(\mathbf{r}_{\mathrm{iso}}(t) - \mathbf{r}_{\mathrm{target}} \right), \tag{14.7}$$

where,

$\mathbf{r}_{\mathrm{iso}}^{\mathrm{comp}}(t)$ is the position of the isocenter in the computational phantom.

$\mathbf{r}_{\mathrm{target}}^{\mathrm{comp}}$ is the position of the target organ in the computational phantom.

$\mathbf{r}_{\mathrm{iso}}(t)$ is the position of the isocenter extracted from RDSR (Eq. 14.5).

$\mathbf{r}_{\mathrm{target}}$ is the position of the target determined by Eq. 14.6.

14.4   RECONSTRUCTING THE SOURCE-TO-SURFACE DISTANCE

The source-to-surface distance (SSD) is an essential parameter for certain radiation dose calculations, such as absorbed skin dose determination for assessing the risk of tissue reactions following a complex image-guided intervention[114]. Yet the SSD is typically not known, as the location of the patient entrance surface relative to the x-ray tube source may vary considerably from one exposure series to the next. The SSD must therefore be estimated by other means. One approach is to approximate the distance from the x-ray source to the patient surface as the distance from the x-ray source to the surface of a phantom model whose size and shape mimics the patient habitus. Given that the computational phantom has been aligned with the patient anatomy (described in the previous section), the SSD can be approximated by finding the line-surface intersection point of a ray line traced from the source to the phantom surface, that is, by analytical ray tracing.

A ray line is for this purpose conveniently expressed as a function of irradiation time t, parametrized in k, with an arbitrary point on the line given by,

$$\mathcal{L}(t) = \mathbf{u}(t) + k\mathbf{v}(t). \tag{14.8}$$

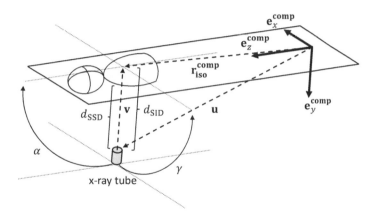

Figure 14.3: The geometry considered for reconstruction of the source-to-surface distance (SSD), d_{SSD}. Shown are relevant rotation angles (α, γ) and position vectors ($\mathbf{u}$, $\mathbf{v}$, $\mathbf{r}_{\text{iso}}^{\text{comp}}$ – the isocenter position) in the computational reference system $\{\mathbf{e}_x^{\text{comp}}, \mathbf{e}_y^{\text{comp}}, \mathbf{e}_z^{\text{comp}}\}$. Also shown is the source-to-isocenter distance (SID), d_{SID}.

- The intercept of the parameterized line is the position of the x-ray source (focal point) in the computational reference system (see Figure 14.3),

$$\mathbf{u}(t) = \mathcal{T}\left(\mathbf{r}_{\text{iso}}^{\text{comp}}(t)\right) \mathcal{R}_z\left(\gamma(t)\right) \mathcal{R}_x\left(\alpha(t)\right) \mathbf{e}_y^{\text{comp}} d_{\text{SID}}(t), \tag{14.9}$$

where d_{SID} is the source-to-isocenter distance (SID), and the different vector transformations are applied as described in Section 14.2.

- The parameterized line extends from the x-ray source in the direction toward the isocenter position, that is,

$$\mathbf{v}(t) = \mathbf{r}_{\text{iso}}^{\text{comp}}(t) - \mathbf{u}(t). \tag{14.10}$$

The vectors that define the ray line expressed by Eq. 14.8 can be calculated as follows:

```
%% Position vectors defining a ray line
% [INPUT]:  RDSR   - data extracted from RDSR, rearranged into a struct.
%           r_iso  - the isocenter position vector relative to the end of
%                    the table (output of Patient_beam_alignment.m script).
%           r_target  - the target position vector relative to the end of
%                    the table (output of Patient_beam_alignment.m script).
%           rc_target - the target position in the computational coordinate
%                    system (example data include in the .m file)
% [OUTPUT]: (u,v)  - position vectors defining a ray line L = u+k*v,
%                    (eq. 14.8), in the computational coordinate system.


% The x-ray beam isocenter position (x;y;z) [cm;cm;cm] in the computational
% coordinate system (r^comp_iso in eq. 14.7), for each irradiation event
% (exposure series) part of a cardiovascular intervention:
rc_iso = rc_target + (r_iso-r_target);


% A ray line can be constructed using RDSR data:
gam  = RDSR.PositionerPrimaryAngle   ; % Rotation angle [rad].
alph = RDSR.PositionerSecondaryAngle ; % Rotation angle [rad].
```

(code continues on next page)

```
dSID = RDSR.DistanceSourcetoIsocenter; % SID [cm].
eyc  = [0;1;0]; % Basis vector in the computational coordinate system.

for i = 1:numel(rc_iso(1,:)) % For each irradiation event.
   R = Rz(gam(i))*Rx(alph(i)); % Rotation (see section 14.2.3).

   u(:,i) = rc_iso(:,i)+R*eyc*dSID(i); % According to eq. 14.9.
   v(:,i) = rc_iso(:,i)-u(:,i);        % According to eq. 14.10.
end
```

The choice of ray-trace algorithm should ideally take into account the type of computational phantom used. For a voxelized phantom model, an iterative (i.e., step-by-step linearized iteration) ray-trace algorithm may be used to solve for which k in Eq. 14.8 the ray line intersects a surface voxel. For a stylized (mathematical) phantom model, an analytical solution can be found for the ray-surface intersection. Since the latter approach can be implemented in a MATLAB code without having to rely on the import of a computational phantom, ray tracing for a stylized phantom has been chosen as a demonstrative example.

The mathematical phantom models considered are the ones used in the Monte Carlo system $PCXMC$[115], which makes use of the slightly modified and updated phantoms of Cristy and Eckerman[116]. The surface of the skull, neck and trunk of these models can be expressed in terms of $(x, y, z) = $ (lateral, vertical, longitudinal) coordinates by the implicit equation of the (quadric) surface of an ellipsoid,

$$\left(\frac{x - x_c}{a}\right)^2 + \left(\frac{y - y_c}{b}\right)^2 + \left(\frac{z - z_c}{c}\right)^2 = 1, \tag{14.11}$$

where the semi-axis lengths (a, b, c) and the center of the geometrical object, $\mathbf{d} = (x_c, y_c, z_c)$, define respectively, the shape and location of an ellipsoid surface or its degenerate, for example, a cylinder surface or a plane surface. The surfaces of the mathematical phantoms are specified in Figure 14.4.

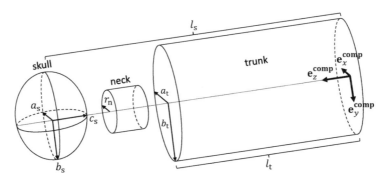

Quadric surface (region)	Semi-axis length	Center	Range
Ellipsoid (skull)	(a_s, b_s, c_s)	$(0, 0, l_s)$	—
Plane (neck top)	$(\to \infty, \to \infty, \to 0)$	$(0, 0, l_s)$	$(x/r_n)^2 + (y/r_n)^2 < 1$
Circular cylinder (neck)	$(r_n, r_n, \to \infty)$	$(0, 0, 0)$	$l_t < z \leq l_s$
Plane (trunk top)	$(\to \infty, \to \infty, \to 0)$	$(0, 0, l_t)$	$(x/a_t)^2 + (y/b_t)^2 < 1$
Elliptic cylinder (trunk)	$(a_t, b_t, \to \infty)$	$(0, 0, 0)$	$0 \leq z \leq l_t$
Plane (trunk base)	$(\to \infty, \to \infty, \to 0)$	$(0, 0, 0)$	$(x/a_t)^2 + (y/b_t)^2 < 1$

Figure 14.4: Simplified representation of the surfaces of the phantoms of Cristy and Eckerman[116], in the computational reference system $\{\mathbf{e}_x^{\text{comp}}, \mathbf{e}_y^{\text{comp}}, \mathbf{e}_z^{\text{comp}}\}$. Semi-axis length and center refers, respectively, to (a, b, c) and $\mathbf{d} = (x_c, y_c, z_c)$ in Eq. 14.11.

Note that the size of the phantoms can be scaled to better correspond to the patient habitus by multiplying the tabulated semi-axis lengths and the center of the geometrical objects with patient specific factors $\mathbf{s} = (s_{xy}, s_{xy}, s_z)$[117]:

$$
\begin{aligned}
s_{xy} &= \sqrt{h_0 m / h m_0}. \\
s_z &= h / h_0.
\end{aligned}
\tag{14.12}
$$

where h and m are the height and weight of the patient, and h_0 and m_0 are the age-dependent nominal height and weight of the phantom model.

The phantom models considered can be included in a MATLAB code as:

```
%% Stylized (mathematical) phantom models
% General characteristics (age [years], height [cm], weigh [kg]):
phantom.age     = [ 0   ,   1  ,    5  ,   10  ,   15  ,   30   ];
phantom.height  = [50.9 , 74.4 , 109.1 , 139.8 , 168.1 , 178.6 ]; % h_0.
phantom.weight  = [ 3.4 ,  9.2 ,  19.0 ,  32.4 ,  56.3 ,  73.2 ]; % m_0.
% Principal dimensions in units of cm (specified in figure 14.4):
phantom.skull.a = [ 4.52,  6.13,   7.13,   7.43,   7.77,   8.00]; % a_s.
phantom.skull.b = [ 5.78,  7.84,   9.05,   9.40,   9.76,  10.00]; % b_s.
phantom.skull.c = [ 3.99,  5.41,   6.31,   6.59,   6.92,   7.15]; % c_s.
phantom.skull.l = [30.17, 42.50,  54.80,  67.18,  83.15,  91.90]; % l_s.
phantom.neck.r  = [ 2.8 ,  3.6 ,   3.8 ,   4.4 ,   5.2 ,   5.4 ]; % r_n.
phantom.trunk.a = [ 5.47,  7.56,   9.82,  11.92,  14.83,  17.20]; % a_t.
phantom.trunk.b = [ 4.9 ,  6.5 ,   7.5 ,   8.4 ,   9.8 ,  10.0 ]; % b_t.
phantom.trunk.l = [21.6 , 30.7 ,  40.8 ,  50.8 ,  63.1 ,  70.0 ]; % l_t.
% Target organ position (cm,cm,cm) (r^comp_target in eq. 14.7):
phantom.heart.x = [-0.02, -0.01,   0.16,   0.12,   0.11,   0.27];
phantom.heart.y = [-1.68, -2.43,  -2.68,  -2.87,  -3.48,  -3.37];
phantom.heart.z = [15.97, 22.30,  29.47,  36.45,  44.92,  49.80];
```

Combining the expression for the ray line (Eq. 14.8) with the expression for the phantom surface (Eq. 14.11), the line-surface intersection can be derived as $\mathcal{L}(k) = \mathbf{u} + k\mathbf{v}$, where,

$$
k = \frac{-(\mathbf{U} \cdot \mathbf{V}) \pm \sqrt{\|\mathbf{V}\|^2 - \|\mathbf{U} \times \mathbf{V}\|^2}}{\|\mathbf{V}\|^2},
\tag{14.13}
$$

with,

$$
\begin{aligned}
\mathbf{U} &= \mathbf{w} . * (\mathbf{u} - \mathbf{d}). \\
\mathbf{V} &= \mathbf{w} . * \mathbf{v}. \\
\mathbf{w} &= (a^{-1}, b^{-1}, c^{-1}).
\end{aligned}
\tag{14.14}
$$

The above equation is a quadratic formula that produces two solutions: k_+ for the case with a positive sign of the root, and k_- for the case with a negative sign of the root. It should be noted that for the present task, complex solutions are possible if there are no intersection points, and more than two solutions are possible if considering multiple surface objects. In case of more than one real-valued solution $(k_1, k_2, ..., k_n)$, the solution that gives the shortest distance from the x-ray source to the line-surface intersection point corresponds to the source-to-surface distance,

$$
d_{\text{SSD}}(t) \equiv \min\left(\left\{ \|\mathcal{L}(k; t) - \mathbf{u}(t)\| \mid k \in \{k_1, k_2, ..., k_n\} \right\}\right).
\tag{14.15}
$$

The described approach can for the exposure of the trunk be implemented as:

```
%% Estimating the SSD for the exposure of a patient's trunk
% [INPUT]:  patient - a struct containting patient information.
%           phantom - a struct containting phantom information.
%           (u,v)   - position vectors defining a ray line L = u+k*v.
%                     (example input data is included in the .m file)
% [OUTPUT]: dSSD    - array with source-to-surface distances.

% An anonymous function (+/-) is constructed for (k+,k-) in eq. 14.13:
kp = @(U,V) - (dot(U,V) + sqrt(norm(V)^2-norm(cross(U,V))^2))/norm(V)^2;
km = @(U,V) - (dot(U,V) - sqrt(norm(V)^2-norm(cross(U,V))^2))/norm(V)^2;

% A phantom model is selected by rounding the patient age to the nearest
% phantom model age [0 1 5 10 15 30] (Patients over 18y are dealt the
% adult phantom (30y)):
modelNo = sum((patient.age./[0.5, 3.0, 7.5, 12.5, 18]) > 1) + 1;
% And patient-specific scaling (sxy,sxy,sz) is applied using eq. 14.12:
sxy = sqrt( phantom.height(modelNo)*patient.weight / ...
            (phantom.weight(modelNo)*patient.height) );
sz  = patient.height/phantom.height(modelNo);
s = [sxy;sxy;sz]; % Scaling vector.

% The vectors U and V that enter eq. 14.13 are calculated:
w = s.*[phantom.trunk.a(modelNo); phantom.trunk.b(modelNo); Inf].^-1;
d = s.*[0                        ; 0                        ; 0 ]   ;
U = w.*(u-d); V = w.*v; % According to eq. 14.14.

for i = 1:numel(u(1,:)) % For each irradiation event.
    % The intersection points are determined according to eq. 14.8:
    Lp = u(:,i) + kp(U(:,i),V(:,i))*v(:,i);
    Lm = u(:,i) + km(U(:,i),V(:,i))*v(:,i);
    % And the SSD is determined according to eq. 14.15:
    dSSD(i) = min( norm(Lp-u(:,i)), norm(Lm-u(:,i)) );
end
```

14.5 CALCULATING THE INCIDENT AIR KERMA

The incident air kerma (K_i) is one of the principal quantities for patient dosimetry in the IAEA TRS-457 formalism for dosimetry in diagnostic radiology [110]. It can be converted into absorbed organ dose using conversion factors or Monte Carlo dose calculations, taking into account the irradiation geometry of each x-ray exposure part of the x-ray procedure [113]. K_i is defined as the air kerma free-in-air (measured or calculated) on the central beam axis at the position of the patient surface. Considering that the exact position of the patient surface is rarely known during an interventional procedure, x-ray systems report the air kerma free-in-air at the patient entrance reference point (K_{ref}); the reference point is conventionally defined as the point on the central beam axis located 15 cm from the isocenter towards the x-ray tube. The incident air kerma can thus be determined from the reference air kerma as,

$$K_i = K_{ref} \, f_{table} \, (d_{SRD}/d_{SSD})^2, \qquad (14.16)$$

where the table transmission factor, f_{table}, accounts for the attenuation of the patient table, d_{SRD} is the source-to-reference distance, and d_{SSD} is the source-to-surface distance. The table transmission can be determined experimentally, or taken into account analytically as described in Ref. [113]. For simplicity, the example MATLAB implementation below uses predetermined table transmission factors.

Having reconstructed the SSD as described in the previous section, K_i can for each separate x-ray exposure be calculated as:

```
%% Calculating the incident air kerma
% [INPUT]:  RDSR   - data extracted from RDSR, rearranged into a struct.
%           dSSD   - array with source-to-surface distances (section 14.4)
%                    (output of Source_to_surface_distance.m script).
%         ftable - array with predetermined table transmission factors
%                    (example input data is included in the .m file).
% [OUTPUT]: Ki     - array with incident air kerma values.

% The source-to-reference distance is determined:
dSRD = RDSR.DistanceSourcetoIsocenter-15; % 15 cm from the isocenter
% The incident air kerma is determined by element-wise multiplication:
Ki   = RDSR.Kref.*ftable.*(dSRD./dSSD).^2; % According to eq. 14.16.
```

14.6 CONCLUSION

This chapter demonstrated how to apply rigid 3D transformations (translation, scaling, rotation) to vectors in MATLAB. Using this elementary mathematics and information sourced from RDSR data, in combination with an algorithm for reconstructing the target organ location, enables the exposure geometry for radiology examinations to be reconstructed. Based on such a framework it is possible to perform dosimetric calculations. Here the estimation of source-to-surface distance and incident air kerma for a set of exposures was demonstrated.

MATLAB toolboxes used in this chapter:			
None			
Index of the in-built MATLAB functions used:			
cat	deal	norm	sin
ceil	dot	numel	sqrt
cos	median	pi	sum
cross	min	repmat	

Simulation of anatomical structure in mammography and breast tomosynthesis using Perlin noise

Magnus Dustler

Diagnostic Radiology, Department of Translational Medicine, Lund University, Skåne University Hospital, Malmö, Sweden

CONTENTS

SOFTWARE PHANTOMS, that is computer models of human anatomy, are useful in many situations, not the least for optimizing radiological imaging parameters without exposing human subjects to ionizing radiation. This chapter describes an attempt to simulate breast tissue, using an approach that can generate breast tissue-like 3D texture using an implementation of so-called Perlin noise.

15.1 INTRODUCTION

15.1.1 Breast imaging

Breast cancer is both the most common cancer among women in industrialized countries and among the most common forms of cancer worldwide with the incidence projected to keep rising due to the aging population [118]. Beginning in the 1970s and 80s, population based screening programs were initiated to facilitate the early detection of breast cancer, and thereby reduce mortality. The dominant screening modality was, and still is, mammography: 2D-breast radiography, today almost exclusively in the form of digital mammography (DM). In recent years, however, a probable successor has emerged in the form of Breast Tomosynthesis (BT), a tomographic, limited angle, imaging method which allows the reconstruction of a 3D breast volume. See Figure 15.1 for example images for both modalities. Several studies show superior sensitivity for BT[119, 120, 121]. Radiographically speaking, the breast consists of just two different types of tissue, adipose (fatty) and fibroglandular (dense), distinguished by their differing x-ray attenuation. Essentially, the breast can thus be considered a binary structure in the radiographic context (with the exception of tiny, highly attenuating microcalcifications).

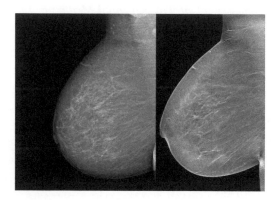

Figure 15.1: Examples of digital mammography (*left*) and breast tomosynthesis (*right*) images of the same breast. Note the visible spiculated tumour on the right image.

As a screening modality, mammography, and its possible replacement breast tomosynthesis, operates under a different set of priorities and constraints than a clinical, diagnostic modality. It needs to be good at finding signs, especially early signs, of cancer while also being very good at correctly identifying healthy women. In other words, the system must operate at a high level of sensitivity while still maintaining a very high specificity. Although, for example, a 95% specificity sounds high, the fact is that because the vast majority of women are healthy, this would for a typical screening population mean close to ten false positives for every detected cancer, as the incidence per screening round is about six per thousand. Studies on breast screening thus normally require very large populations to reach sufficient statistical power. This is especially true of innovations in image processing and (for breast tomosynthesis) reconstruction, which are not normally expected to lead to dramatic differences in detection performances. Such studies require large investments in both time and money, and may also involve subjecting thousands of healthy women to ionizing radiation. Thus, an alternative to human subjects is highly attractive, which leads to the issue of simulating sufficiently realistic breast tissue.

15.1.2 Simulation of breast tissue

Several different groups have investigated various approaches to simulating breast tissue, and have done so for different reasons. The simplest way is to not simulate at all, but to instead employ existing patient images and modify them, for example by adding circular features meant to represent solid lesions. There are several pitfalls to this method, the most serious of which is that the breast itself, of course, is a 3D-structure, and, at least in the DM case, we lack information about the 3D location of different structures in the breast. This only slightly improves with BT, as it cannot provide reliable Hounsfield unit values for the individual voxels owing to the limited angular range. Perhaps the most important advantage of instead using software models is the exact control of ground truth, with the true contents of each voxel known from the outset. This allows the location of simulated lesions to be precisely determined, and for the exact material properties to be set. This may assume that a model of the imaging equipment is also available, that can generate a synthetic image from a software phantom. One could argue that this is not fully realistic, but in many situations, the advantages outweigh the disadvantages. The other major benefit of software models is flexibility. While the appearance and properties of a physical phantom is more or less fixed, a software phantom can, with the right programming, be set up in an unlimited number of configurations and variations.

15.1.3 Basic breast tissue appearance

Breast tissue consists mainly of fibrous connective tissue, blood vessels, milk ducts and milk glands and fat (adipose) tissue. The tissue components are varied in distribution and appearance, making it difficult to make general statements on radiographic appearance of breasts. There are several ways of categorizing the amount and distribution of dense tissue, the most widely used of which is the BI-RADS [122] density classification, which in its latest version classifies breasts in four categories, ranging from A which is almost completely adipose, to D which is almost fully and homogenously dense. The B and C categories contain breast dominated by smaller clumps and fibrous strands of dense tissue. See Figure 15.2.

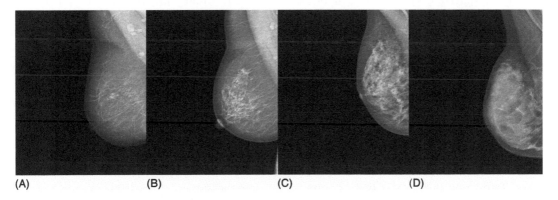

(A) (B) (C) (D)

Figure 15.2: Examples of BI-RADS breast density categories from A (*least dense*) on the left, to D (*most dense*) on the right.

Another important feature of breast appearance is of course its shape and outline, and also how this shape is affected by the angle of the projection and the compression during mammography. This will not be discussed here, as it is quite a different problem from the texture of the tissue and would require a lengthy departure into the field of finite element analysis and deformable meshes. Interested readers should refer to e.g. [123].

The spatial frequency distribution of breast structures is known to more or less follow a power law distribution with an exponent of 2-3[124]. This has led to a fast and simple approximation of breast tissue by simply generating 2D power law noise, which can look rather similar to an actual breast (see Figure 15.3). However, this realization lacks the continuous structures and connections seen in real breasts. More complex methods are needed.

Figure 15.3: A 2D projection of power law noise.

15.1.4 Perlin noise

The algorithm properly known as Noise, was first described by Ken Perlin in 1985 [125]. It has since become very common in the computer graphics and special effects industries. It is a flexible algorithm which generates smoothly and continuously varying structures of a set spatial frequency. It can be realized in any dimension, but for the purposes of this chapter we are interested in a three-dimensional implementation, which is normally visualized by projecting the volume into a two-dimensional image. However, in order to make it easier to grasp the basic implementations of the algorithm we will sometimes use two-dimensional examples in the following section to illustrate concepts, before extending to three dimensions.

15.2 GENERATING THE NOISE

15.2.1 Gradient fields

The most basic component of Perlin noise is the *gradient field*. It defines the appearance and frequency (size scale) of the noise, and using the same field will always generate the same finished noise volume. First we will consider a simple example in two-dimensions, a gradient field of the minimum possible size (see Figure 15.4).

Consider a unit square. Each corner of the square is assigned one random two-dimensional vector $\mathbf{g}$, normalized to have a length of one. This can be seen as the gradient field for Perlin noise of, somewhat arbitrarily, frequency 1–or 1×1. To increase frequency, add more unit squares with associated gradient vectors, 2×2, 3×3 etc. In the three-dimensional case, we use unit cubes instead, $1 \times 1 \times 1$, $2 \times 2 \times 2$ etc. Using MATLAB, a gradient field in three dimensions can be generated as follows:

```
function [gradients] = randomGradfield(inSize)
% Generates a random set of gradient vectors associated with a 3D gradient
% field
% inSize: Size of gradient field in a 1x3 vector [i,j,k]
% gradients: nX3 set of gradient coordinates
[x,y,z] = cube12rand((inSize(1)+1)*(inSize(2)+1)*(inSize(3)+1));
gradients = single([x' y' z']');
end
```

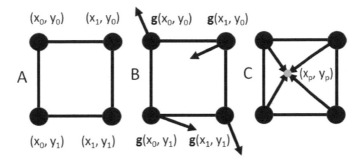

Figure 15.4: The basic theory behind 2D Perlin noise. The gradient field consists of a number of gradient squares, one of which is shown in A. Each corner of the square is assigned a random gradient vector, shown in B. For each point on which the noise function will be evaluated, the grid, find the distance vectors from the four closest corners, as shown in C.

In this function, `inSize` represents the number of cubes along each dimension. We will generally consider only isotropic volumes (i.e., $n \times n \times n$ cubes), though anisotropic volumes can also be used. The **cube12rand** function generates the gradient vectors. Note that it is possible to use any number of different methods for generating pseudorandom gradient vectors. It is not vitally important that the vectors are truly random, but it is important that they are uniformly distributed across the surface of a unit sphere and that there is no clustering along the principal axes, as this will introduce biases in the appearance of the noise. Be aware that simply using repeated calls to `rand(1,3)` will not generate appropriate vectors, as they will be distributed along the surface of a unit cube rather than unit sphere and will introduce a blocky appearance (which could be interesting for certain applications). One can of course design an algorithm that generates vectors with an appropriately random, spherical distribution, but it does not improve the appearance of the noise compared to drawing them from a finite set of uniformly distributed ones. The method proposed by Ken Perlin is a fast pseudorandom sampling of the 12 vectors from the center of a cube (with its center at [0,0,0]) to its edges. It can be implemented like this:

```
function [ex,ey,ez] = cube12rand(n)
% Generates [i,j,k] coordinates of n vectors chosen from the set defined
% below
gradients = [[0 1 1]' [0 1 -1]' [0 -1 1]' [0 -1 -1]' [1 0 1]' [-1 0 1]'...
    [1 0 -1]' [-1 0 -1]' [1 1 0]' [-1 1 0]' [1 -1 0]' [-1 -1 0]'];
gradients = bsxfun(@rdivide, gradients, sqrt(sum(abs(gradients).^2,1)));
numbers = randi([1 12],1,n);
ee = gradients(:,numbers);
ex = ee(1,:);
ey = ee(2,:);
ez = ee(3,:);
end
```

15.2.2 The grid

The *grid* is a rather fanciful name for the set of points on which the Perlin noise function is evaluated. It is independent of the size of the gradient field (though it must be larger) and determines the size or resolution of the generated noise volume. Again considering the two-dimensional case for simplicity, if we use a 200×200 grid over a 2×2 set of gradient squares, there will be 100×100 pixels in each square. If we use a 2000×2000 grid, we will increase the resolution to 1000×1000 pixels per square.

To easily generate the coordinates in a useful format, the built-in `meshgrid` function can be used, with the input variable `gridSize` being the desired size of the grid:

```
function [x,y,z,gridSize] = generateGrid(gridSize)
% Generate grid
gridSize = single(gridSize);

if length(gridSize) == 1
    [X,Y,Z] = meshgrid(1:gridSize,1:gridSize,1:gridSize);
    gridSize = [gridSize gridSize gridSize];
else
    [X,Y,Z] = meshgrid(1:gridSize(1),1:gridSize(2),1:gridSize(3));
end

% Reshape coordinate gridSizes to vectors
x = reshape(single(X),gridSize(1)*gridSize(2)*gridSize(3),1);
y = reshape(single(Y),gridSize(1)*gridSize(2)*gridSize(3),1);
z = reshape(single(Z),gridSize(1)*gridSize(2)*gridSize(3),1);
end
```

15.2.3 Generating noise

Describing the next step is relatively straight forward, though it takes some effort to code and compute. A caveat: the implementation used here works and is reasonably easy to understand, but is not necessarily the most efficient one.

We have the gradients and grid points, which gives us all the building blocks for the next step, which is to evaluate the Perlin noise function for each point `G(i)` (or `G(x,y,z)`). Once again, first consider the 2D step. For each point, find the four closest corners, that is, the corners of the square it is contained within. Now, find the four position vectors between the point and the corners and calculate the dot (scalar) products of these four vectors and the corresponding gradients of the matching corners. As the final step, take a weighted average of these products and you will have the noise value of the point.

15.2.4 Finding corners

The following section of code shows how to find the eight cube corners associated with each point in the grid. The variable `fieldSize` is a vector defining the size of the gradient field along each dimension, that is, $(3,3,3)$ would mean a cube of 27 unit cubes:

```
% Find size of cubes in Perlin field
cubeSize = min(gridSize)./min(fieldSize);
% Number of edges to gradient field
fieldSizePlus1 = fieldSize+1;
```

(code continues on next page)

```
% Determine position of points within field cube
x1 = ceil(x/cubeSize)+1;
x0 = x1 - 1;
y1 = ceil(y/cubeSize)+1;
y0 = y1 - 1;
z1 = ceil(z/cubeSize)+1;
z0 = z1 - 1;

% Find gradients of cube corners for each point
% 1: [0 0 0] 2: [0 1 0] 3: [1 0 0] 4: [1 1 0]
% 5: [0 0 1] 6: [0 1 1] 7: [1 0 1] 8: [1 1 1]
grad = single(zeros(length(x),24)); % Preallocate matrix
% Gradient of 1st corner
grad(:,1:3) = gradients(:,sub2ind([fieldSizePlus1(1) fieldSizePlus1(2) ...
    fieldSizePlus1(3)],x0,y0,z0))';
% Gradient of 2nd corner
grad(:,4:6) = gradients(:,sub2ind([fieldSizePlus1(1) fieldSizePlus1(2) ...
    fieldSizePlus1(3)],x0,y1,z0))';
% Gradient of 3rd corner
grad(:,7:9) = gradients(:,sub2ind([fieldSizePlus1(1) fieldSizePlus1(2) ...
    fieldSizePlus1(3)],x1,y0,z0))';
% Gradient of 4th corner
grad(:,10:12) = gradients(:,sub2ind([fieldSizePlus1(1) ...
    fieldSizePlus1(2) fieldSizePlus1(3)],x1,y1,z0))';
% Gradient of 5th corner
grad(:,13:15) = gradients(:,sub2ind([fieldSizePlus1(1) ...
    fieldSizePlus1(2) fieldSizePlus1(3)],x0,y0,z1))';
% Gradient of 6th corner
grad(:,16:18) = gradients(:,sub2ind([fieldSizePlus1(1) ...
    fieldSizePlus1(2) fieldSizePlus1(3)],x0,y1,z1))';
% Gradient of 7th corner
grad(:,19:21) = gradients(:,sub2ind([fieldSizePlus1(1) ...
    fieldSizePlus1(2) fieldSizePlus1(3)],x1,y0,z1))';
% Gradient of 8th corner
grad(:,22:24) = gradients(:,sub2ind([fieldSizePlus1(1) ...
    fieldSizePlus1(2) fieldSizePlus1(3)],x1,y1,z1))';

% Determine relative coordinates of points within relevant cubes
xrel = (x/cubeSize+1)-x0;
yrel = (y/cubeSize+1)-y0;
zrel = (z/cubeSize+1)-z0;

% Corners: Relative positions
% 1: [0 0 0] 2: [0 1 0] 3: [1 0 0] 4: [1 1 0]
% 5: [0 0 1] 6: [0 1 1] 7: [1 0 1] 8: [1 1 1]
c0 = single(zeros(length(x),1));
c1 = single(ones(length(x),1));
corners = {[c0 c0 c0] [c0 c1 c0] [c1 c0 c0] [c1 c1 c0] [c0 c0 c1]...
    [c0 c1 c1] [c1 c0 c1] [c1 c1 c1]};
```

15.2.5 Vectors and gradients

The next step is to take the position vectors, multiply them by the correct gradients and then calculate the weighted average for each point. The first part consists of finding the sum of the dot products, which is simply multiplying the relevant parts of the variable **grad** for each of the eight corners with the corresponding position vectors, and summing the results.

The position vectors are constructed by subtracting the relative locations of the corners (the **corners** vector) from the relative x,y and z coordinates of each point (see Figure 15.4c).

```
% Calculate vectors between points and corners and calculate dot products
% between them and respective gradients, i.e., calculate corner values 1:8
dots = single(zeros(length(x),8));
for i = 1:8
    dots(:,i) = sum(grad(:,(1+3*(i-1)):(3+3*(i-1)))'.*([xrel yrel zrel]'...
        -corners{i}'))';
end
% Perform weighted averaging
[v] = weightedAverage(xrel,yrel,zrel,dots);
```

Now, we need to take the averages of the eight values for each point. This sounds very straight forward and would normally involve nothing more than a call to the **mean** function. But for Perlin noise, it is highly recommended to use a weighted average instead, a so-called fade or smooth-step function. This is to smooth out the averages so that high values tend towards 1, low values tend towards zero and middle values tend towards 0.5. This might sound odd, but it has been found that to the human eye, the generated noise looks more natural this way. For this we use the following:

```
function [v] = weightedAverage(xrel,yrel,zrel,dots)
% Distances
tx0 = xrel;
tx1 = abs(xrel-1);
ty0 = yrel;
ty1 = abs(yrel-1);
tz0 = zrel;
tz1 = abs(zrel-1);

% Define weighted average polynomial
w = @(x)6.*x.^5-15.*x.^4 + 10.*x.^3;

% Calculate weights
wx0 = w(tx0);
wx1 = w(tx1);
wy0 = w(ty0);
wy1 = w(ty1);
wz0 = w(tz0);
wz1 = w(tz1);

% Weights of corners 1:8
weight(:,1) = wx1.*wy1.*wz1;
weight(:,2) = wx1.*wy0.*wz1;
weight(:,3) = wx0.*wy1.*wz1;
weight(:,4) = wx0.*wy0.*wz1;
weight(:,5) = wx1.*wy1.*wz0;
weight(:,6) = wx1.*wy0.*wz0;
weight(:,7) = wx0.*wy1.*wz0;
weight(:,8) = wx0.*wy0.*wz0;

% Weighted sum
v = sum(dots.*weight,2);
end
```

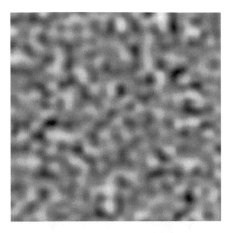

Figure 15.5: A 2D projection of a basic Perlin noise simulation.

The function is used to assign weights based on the proximity of the evaluated point to 0 and 1, along each axis. These are then combined to calculate weights for each corner which are multiplied with the corresponding dot products. The alternative is to do a linear or cubic interpolation between the dot products.

The very last step is to collect and reshape the results in readable manner.

```
% Collect results
inds = sub2ind([gridSize(1) gridSize(2) gridSize(3)],x,y,z);
out = zeros(gridSize(1),gridSize(2),gridSize(3),'single');
out(inds) = v;
```

This results in 3D noise which looks (depending on frequency) like Figure 15.5 when projected into the plane. We have collected the code presented above into a function **perlinNoise(fieldSizeSc,gridSize)**, which is provided to the reader along with this chapter's other functions and scripts (see the book's accompanying code repository). The argument **fieldSizeSc** is a scalar value defining the size of the gradient field (in every dimension); **gridSize** is the size of the grid, and can either be passed as a scalar (defining equal size in each dimension) or a 1x3 vector. An image equivalent to Figure 15.5 can be generated as follows:

```
% Generate 3D Perlin noise
[noise] = perlinNoise(16,256);

% Project to a 2D noise image
noise2D - mean(noise,3);

% Display the image
imshow(noise2D,[])
```

The noise forms visually continuous structures with intensity values smoothly varying between 1 and -1, with few hard transitions.

We now have the tools to generate 3D Perlin noise of any spatial frequency, but as we will see in the next section, there are quite a few steps left until we have a realistic-looking volume of breast tissue.

15.3 FRACTAL NOISE

While Perlin noise as described in the preceding section provides interesting structures, its structures are band-limited to a narrow range of spatial frequencies. As most real structures have a much wider frequency range, we must be able to combine structures of different frequencies into a convincing composite.

15.3.1 Frequencies of noise

The frequency of Perlin noise can easily be modified by varying the size of the gradient field. This is shown in Figure 15.6 by a collage of images produced by calls to the `perlinNoise` function with an increasing field size.

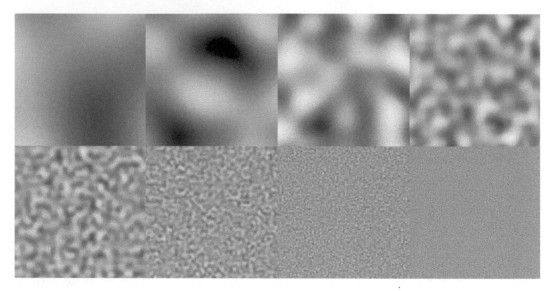

Figure 15.6: A collage of Perlin noise images of low to high frequency.

Note that if the gradient field size approaches the size of the grid the output will just be random noise.

```
for index = 1:8
[out{index}] = perlinNoise(2^(index-1),256);
end
```

Another way of doing this is to generate a single high frequency gradient field and zoom in, using a successively smaller part of it for the lower frequencies. This nicely illustrates the fractal properties of Perlin noise, as each part exhibits similarity to the whole.

15.3.2 Building up the noise: lacunarity and persistence

To generate Perlin noise with a wider range of frequencies, we use a method where noise is generated at successively higher frequencies and added together to form a volume with a range of differently scaled structures. The properties of each successive noise volume—known as an octave in analogy with musical terminology—is defined by two parameters known as *lacunarity* and *persistence*. Lacunarity defines the frequency gap between successive octaves: a constant that multiplies the frequency of the preceding octave. Persistence defines the dampening of the amplitude of higher frequencies and is likewise a constant that is multiplied to the value of the noise. By varying these parameters, the characteristics of the noise changes. This process of building up fractal noise can be applied to any noise-generating algorithm, not just Perlin noise but also to, for example, the related Worley noise[126]. As the process is mathematically very similar to Brownian motion, this method of generating noise with fractal properties is often called fractional or fractal Brownian motion. See below for an example of code that runs using the `perlinNoise` base function:

```
function [noise,octs] = fractalNoise(persistence,lacunarity,gridSize)
% persistence: Amplitude of next octave as fraction of preceeding one
% lacunarity: Frequency of next octave as fraction of preceeding one >1
% gridSize: Size of output noise volume
% Calculate number of octaves needed
nOct = ceil(log(gridSize)/log(lacunarity));

% Run generation loop
noise = zeros(gridSize,gridSize,gridSize);
amplitude = 1;
for iOct = 1:nOct
    frequency = round(lacunarity^(iOct-1));
    octave = amplitude*(perlinNoise(frequency,gridSize));
    noise = noise + octave;
    if nargout == 2
        octs{iOct} = octave;
    end
    amplitude = amplitude*persistence;
end
end
```

This function will generate Perlin noise with a range of frequencies. The sample noise images in Figure 15.6 can for example be generated as the octaves of the following function call:

```
[noise,octs] = fractalNoise(1,2,256);
```

To get a more interesting look, however, it is better to use a lower persistence of say 0.5, so that the higher frequencies are more dampened. The following call:

```
[noise,octs] = fractalNoise(0.5,2,256);
```

will generate the noise shown in Figure 15.7 when the octaves are summed. While this would probably work for simulating clouds, we are not quite there yet in regards to breast tissue.

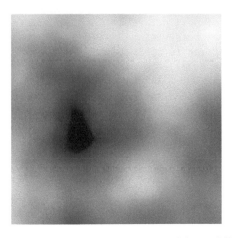

Figure 15.7: An example 2D projection of fractal Perlin noise.

15.3.3 The perturbation function

To add even more interest to the fractal noise, we can add a so-called perturbation or turbulence function, either to the complete volume or to each individual octave before combining them. This turbulence function can be anything, and it is in this step that creating fractal Perlin noise passes from science to art, in a sense, as there is no exact way of finding optimal functions. The main issue here is the look of the noise, not its mathematical underpinning. Perlin noise with different turbulence functions have been used to represent everything from sea-water to fire, but the limited space and scope of this chapter will not allow us to explore these concepts in general. Instead, a few examples that might be interesting will be shown when we continue on to generating actual organic structures. Before that, we will introduce a way of making the code run more effectively.

15.4 PRE-GENERATION

A problem with the implementation is the RAM memory requirement and the time that it needs to run. To run the code of this chapter with a `gridSize` of 256, we recommend that the reader has at least 8 GB of RAM on their computer. This is not unreasonably high. However, as the number of noise realizations (or finished image volumes) simulated is increased, the implementation eventually becomes impractical in terms of execution time. One approach to solving this issue is to pre-generate a set of octaves with appropriate frequencies that can be combined as needed to create finished volumes. This is easy to code, where we can simply pregenerate octaves starting at frequency one and then increase with a desired lacunarity. There is of course no need to generate volumes with different persistence, as the octaves can simply be multiplied by a constant. By generating a number of iterations of each octave, repetition will not be apparent in finished volumes. To further increase the variation, the octaves can also be randomly rotated. This means that with, for example, 8 octaves of noise and 6 iterations of each octave, we can create 6^8 combinations—and many more if we consider random rotations of the volumes.

The following code creates fractal Perlin noise volumes, loading a randomized set of pregenerated octaves to build up the final noise volume. The pregenerated octaves need to be generated before using this function. For convenience, a script **pregenData.m** is provided with this chapter's code for that purpose.

```
function [noise,octs] = fractalPregen(persistence,lacunarity,gridSize,func)
% persistence: Amplitude of next octave as fraction of preceeding one
% (dampening, gain)
% lacunarity: Frequency of next octave as fraction of preceeding one
% gridSize: Size of output noise volume
% func: Perturbation function to apply to octaves
%   eg. @(x)max(max(max(abs(x))))-abs(x);

% Unless a perturbation function specified, apply no perturbation
if nargin <4
    func = @(x) x;
end

% Locate pregenerated data and determine the number of iterarions available
pregenDir = fullfile('pregenData', ...
    ['l' num2str(lacunarity) '_s' num2str(gridSize)]);
nFiles = length(dir(fullfile(pregenDir,'f1_*.mat')));
```

(code continues on next page)

```
if nFiles == 0
    error('Pregenerated data is missing')
else
    nIter = nFiles;

% Calculate number of octaves needed
nOct = ceil(log(gridSize)/log(lacunarity));

% Predefine noise volume
noise = single(zeros(gridSize,gridSize,gridSize));

% Run loop through frequencies
amplitude = 1;
for iOct = 1:nOct
    frequency = round(lacunarity^(iOct-1));
    iIter = randi(nIter); % Randomly select an iteration
    pregenFile = fullfile(pregenDir,['f' num2str(frequency) '_s' ...
        num2str(gridSize) '_i' num2str(iIter) '.mat']);
    vars = load(pregenFile);   % Read in the selected pregenerated octave
    oct = amplitude.*func(vars.oct);
    noise = noise + oct;
    if nargout == 2
        octs{iOct} = oct;
    end
    amplitude = amplitude*persistence;
end
end
```

15.5   THE FINAL TISSUE MODEL

The issue of simulating breast tissue itself is important, but it is complicated by the fact that it varies so widely in appearance. We will not solve this here, but I will show some ideas on where we should go. There are two main areas to overcome: the fibrous, strand-like tissue and the nebulous density. Similar methods can be used for both, but as the culmination of this chapter we will work with an example that focuses on the fibrous strands.

As we discovered earlier, Perlin noise is smoothly varying and continuous. This is often a nice property, but sometimes we are interested in more abrupt transitions, for example to represent fibrous strands in the breast tissue. One way of doing this is to use the absolute value of the noise. As values normally vary smoothly from 1 to -1, this will change the characteristics of the noise so that we retain the smooth variations from 1 to 0, but then get an abrupt switch at 0 before the values rise back to 1. These sharp canyons or valleys easily become ridges instead by inverting the noise values. These ridges look very similar to the strands seen in mammograms. To further enhance the look one can narrow the strands and make the transitions more abrupt. This could be done in different ways, for example, with an exponential function, but here we will raise it by, say, the sixth power. Putting this together gives us the simple code below:

```
[noise] = fractalPregen(0.8,1.5,256);
perturbation = @(x) (1-abs(x)).^6;
perturbedNoise = perturbation(noise);
```

This leads to a volume very similar to the one in Figure 15.8. Compare this volume with the simple simulation in Figure 15.3 and in particular to the examples of real breast tissue in Figures 15.1 and 15.2. For further reading, please see [127] and [128].

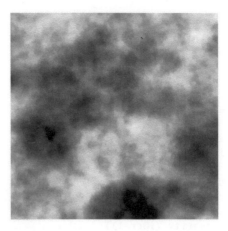

Figure 15.8: A 2D projection of a fractal Perlin noise volume with the addition of a perturbation function, in this case the inverted absolute value to sixth power. This has several characteristics of fibrous breast tissue.

15.6 CONCLUSION: GENERATING BREAST TISSUE

Finally, we are at the end of this chapter and have at our disposal some new tools that can be adapted to a wide variety of situations. Hopefully, this small contribution can inspire others working with similar problems to make their own modifications to lacunarity, persistence and perturbation functions to tune the look of the noise to match the characteristics of the structures they are attempting to simulate. We have combined everything we have learnt and created a Perlin noise volume that looks something like breast tissue, or at least close enough to give an idea of how to proceed.

With the tools provided here there is an infinite number of combinations of lacunarity, persistence and perturbation functions that can be used to adapt Perlin noise to a wide range of tasks and applications in medical imaging. But they are only tools; what is needed now is curiosity and a large dose of trial-and-error.

MATLAB toolboxes used in this chapter:
None

Index of the in-built MATLAB functions used:

abs	length	nargout	sqrt
bsxfun	log	num2str	sub2ind
ceil	load	ones	sum
dir	mean	randi	zeros
error	meshgrid	reshape	
fullfile	min	round	
imshow	nargin	single	

xrTk: a MATLAB toolkit for x-ray physics calculations

Tomi F. Nano and Ian A. Cunningham

Robarts Research Institute and Department of Medical Biophysics, Western University, London, Canada

CONTENTS

M EDICAL x-ray imaging requires images of sufficiently high quality obtained using low levels of radiation exposure, to ensure patients receive the benefits of high-quality medical care with low acceptable risks from radiation exposure. As illustrated in Figure 16.1, image quality and the ability to visualize fine structures is a balance between many factors, including image contrast and image signal-to-noise ratio (SNR). One important task is the visualization of iodinated contrast agents used to make arteries visible in angiography. This chapter describes an application of the xrTk toolkit to calculate both contrast and SNR of iodinated arteries and an optimization of exposure technique to ensure acceptable patient dose.

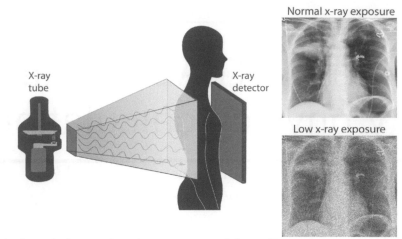

Figure 16.1: A typical x-ray imaging system used for radiography consists of an x-ray tube that produces a diverging x-ray beam and a detector to receive the x-ray intensity transmitted through a patient. Image SNR is dependent on exposure level as illustrated (simulated).

16.1 INTRODUCTION

16.1.1 The toolkit

The MATLAB toolkit xrTk is an open-source library of functions useful for physics calculations in the design and understanding of x-ray imaging systems. It allows the user to: i) generate x-ray spectra used in radiography and mammography; ii) generate x-ray interaction coefficients at specified x-ray energies for stable elements; and iii) perform a variety of physics calculations including those of x-ray attenuation, image contrast, patient exposure, air KERMA and dose. Source code for xrTk is available for download from a git repository[129] under the GPL[130] open-source license. It is different from other libraries[131, 132] as it supports both radiographic and mammographic x-ray spectra and uses arbitrary energy-bin width in all calculations.

16.1.2 Toolkit functions

Table 16.1 lists selected xrTk functions useful for calculations of x-ray spectra, contrast, SNR and dose. In total, xrTk includes 32 functions and uses databases from Hubbell[133] for material properties (older interaction coefficients are also available), NIST[134] for attenuation and absorption coefficients, and XCOM[133] for interaction cross-sections. Properties of pure elements and selected materials such as muscle and bone, including mass and atomic and electronic densities at standard temperature and pressure, are provided by **xrMaterialInfo** for elements with atomic numbers 1 to 99. For specified energies between 1 keV and 100 MeV, **xrCoef** returns a vector of x-ray coefficients for a specified element or material name and **xrCoefFormula** returns coefficients for a specified chemical formula (such as "Gd2O2S"). X-ray spectra are generated by **xrSpec** and **xrSpecIec** using the Tucker-Barnes algorithms[135, 136] for arbitrary Mo or W target spectra and the RQA series of standard spectra[137, 138], respectively. Most other functions are used for manipulating spectra (e.g., **xrTrans** and **xrTransFormula**) and calculating spectra properties (e.g., **xrAirKerma**, **xrExposure**, **xrDose** and **xrHvl**). Many functions make use of variable-length input argument lists (**varargin**) to provide reasonable default values where appropriate, while maintaining a simple structure for passing additional arguments where

Table 16.1: List of selected xrTk functions.

Function	Description
xrMaterialInfo	Material information (such as mass density)
xrCoef	X-ray interaction coefficients and cross-sections
xrGetBin	Energy bin associated with specified energy
xrQpGye	Photons/mm^2/Gy values at specified energies
xrQpRe	Photons/mm^2/R values at specified energies
xrGypQe	Gy/(photons/mm^2) values at specified energies
xrRpQe	R/(photons/mm^2) values at specified energies
xrTrans	X-ray transmission factors at specified energies
xrAirKerma	Air KERMA (Gy) of specified spectrum
xrDose	Absorbed dose (Gy) for a spectrum in a specified material
xrExposure	Exposure (R) of a specified spectrum
xrFMed	F-medium values, ratio of Gy / R, at specified energies
xrHvl	First half-value layer (mm Al) of a specified spectrum
xrSpec	Generates an x-ray spectrum for a molybdonum or tungsten target tube using Tucker-Barnes algorithms
xrSpecIec	Generates standard IEC spectra with HVL matched to IEC values and Q_o closely matched
xrSpectrum2Bins	Converts energy and spectrum vectors to specified bins

desired. Energy spectra are represented as vectors of uniformly-spaced energy values with the spacing selection arbitrary. Small energy spacings can be used when fine detail in interaction coefficients near transition edge energies is required, and larger spacings for faster calculations when this is not required. It is recommended that photon spectra be plotted using **xrSpectrum2Bins** which generates vectors of energy and spectra to plot as bins (stairs) using MATLAB plot functions.

16.1.3 Toolkit validation

It is often difficult to accurately calculate the spectral shape for a particular situation because spectral shape is affected by many factors, some of which may not be known, such as target angle and surface condition (smooth or pitted), tube window material and thickness, and possibly other attenuators such as tungsten plating on the window. For this reason it is more convenient to characterize a spectrum in terms of the kV setting (tube potential), half-value layer (HVL)–which is the thickness of aluminum required to attenuate beam exposure by 50%–and the number of photons per unit area per air-KERMA (Q_o) as defined in Section 16.2.3.

Standard spectra suggested by the International Electrotechnical Commission (IEC)[137] are often used in evaluating detector performance and image quality. Table 16.2 summarizes HVL and Q_o values calculated using xrTk. To accomodate unknown specifics of added or

Table 16.2: Comparison of first HVL and number of quanta per unit air KERMA using xrTk spectra with standard spectra (IEC 62220-1-1 for radiography and IEC 62220-2 for mammography).

Spectrum Name	IEC		xrTk	
	HVL (mm Al)	Q_o (q/mm^2/µGy)	HVL (mm Al)	Q_o (q/mm^2/µGy)
RQA-3 (50 kV, 10 mm Al)	3.8	20,673	3.8	20766 (+0.4%)
RQA-5 (70 kV, 21 mm Al)	6.8	29,653	6.8	29733 (+0.2%)
RQA-7 (90 kV, 30 mm Al)	9.2	32,490	9.2	32644 (+0.5%)
RQA-M3 (30 kV, 2 mm Al)	0.62	5,303	0.62	5450 (+2.8%)
RQA-M4 (35 kV, 2 mm Al)	0.68	6,325	0.68	6275 (-0.8%)
MoRh (28 kV, 2 mm Al)	0.65	5,439	0.65	5511 (+1.3%)

inherent filtration for each spectrum, xrTk assumes 1.5 mm added Al filtration for radiography and 0.032 mm Mo for mammography spectra, and applies a small filtration adjustment to match HVL values. Excellent agreement of Q_o with IEC values was obtained, to within 1% for radiography and 3% for mammography spectra (brackets in right-hand column).

16.2 OPTIMIZING IMAGE QUALITY

16.2.1 X-Ray spectra

X-ray tubes generate x-ray photons by directing a focused high-energy electron beam on to a metal target, creating bremsstrahlung radiation (having a wide spectrum of energies) and characteristic emissions (having a series of specific energies characteristic of the target material). The spectrum, $q(E)$ [quanta mm^{-2}keV^{-1}], describes the number of quanta per unit area and energy as a function of energy E [keV]. The total fluence $\mathcal{N}$ [quanta mm^{-2}], is given by the integral

$$\mathcal{N} = \int_0^{kV} q(E)\mathrm{d}E, \qquad (16.1)$$

where kV is the applied tube potential. This can also be expressed in terms of a summation of the fluence per energy bin,

$$\mathcal{N} = \sum_i q_i. \qquad (16.2)$$

In xrTk, x-ray spectra are represented as vectors where each element q_i [mm^{-2}] is associated with an energy-vector element E_i [keV] on uniform spacings ΔE and gives the number of x-ray quanta per mm^2 having energy in $E_i - \Delta E/2 \leq E_i < E_i + \Delta E/2$. Unlike some other libraries, q_i gives the number of quanta in the energy bin, rather than the number per unit energy, to preserve the area of characteristic emission peaks for any specified energy bin

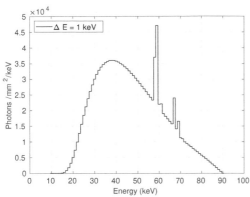 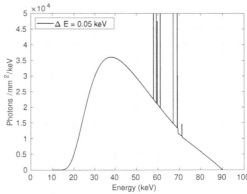

Figure 16.2: X-ray spectra are represented as vectors (q_i) where each element value represents the number of photons per mm^2 in an associated energy-vector bin of width ΔE. For convenience of display, q_i has been divided by the bin width in the plots. Characteristic emission lines appear to have width ΔE and area equal to the total number of photons in the emission line as illustrated for $\Delta E = 1$ keV (*left*) and 0.05 keV (*right*).

width. For this reason it is recommended that **xrSpectrum2Bins** be used to plot spectra as this will preserve bin width, as illustrated in Fig. 16.2.

Spectra are generated using the semi-empirical Tucker-Barnes[135, 136] algorithms for radiography and mammography, at arbitrary energy resolution, using **xrSpec**. The following code segment generates a 90-kV spectrum at 1 m for a 1 mAs exposure technique, and finds the corresponding HVL [mm of Al], air KERMA [µGy], and Q_o [quanta mm^{-2} µGy^{-1}] values:

```
% Generation of x-ray spectra

dE = 0.25; EMin = 10; EMax = 120; % keV
E = EMin:dE:EMax;

kV = 90; target = 'W'; % 90 kV, tungsten target (radiography)
qi = xrSpec(E, kV, target);
hvl = xrHvl(E, qi);
K = xrAirKerma(E, qi)*1e6;
Qo = sum(qi)/K;
```

Values for arbitrary mAs, focus-to-detector distance [m], target angle [degrees], oil thickness and window composition can be specified with optional arguments. Selected IEC[137] spectra can be generated using **xrSpecIec**, as illustrated below:

```
% Generation of standard IEC x-ray spectra

E = 20:.25:100;
spectrumName = 'RQA-5'; iecVersion = 2015;
qi = xrSpecIec(E, spectrumName, iecVersion);
hvl = xrHvl(E, qi);
K = xrAirKerma(E, qi)*1e6;
Qo = sum(qi)/K;
```

This code segment returns a spectrum with HVL matched to the IEC value, and calculates Q_o (see Table 16.2). The **iecVersion** argument can take values '2003' or '2015' for radiographic spectra.

16.2.2   X-Ray interaction coefficients

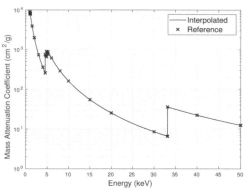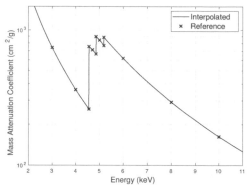

Figure 16.3: X-ray coefficients are determined from reference tables[134] using a log-log cubic-spline interpolation with discontinuities at absorption edges where required. In this example for iodine, cross marks indicate tabulated reference data and solid lines indicate interpolated points on uniform 0.01 keV spacings.

X-ray transmission through an attenuator is described by energy-dependent interaction coefficients. Tabulated coefficients were sourced from the US National Institute of Standards and Technology (NIST) website[134]. Values are interpolated with a log-log cubic-spline interpolation, as recommended by NIST, for any x-ray energy between 1 keV and 100 MeV, using the function `xrCoef`. For example:

```
% Interpolation of x-ray interaction coefficients

E = 1:.01:50;
material = 'iodine';
mu = xrCoef(E, material, 'attenuation'); % Mass attenuation coefficient of iodine
```

This example generates and returns a vector (`mu`) of mass attenuation coefficient values (μ/ρ [cm^{2}g^{-1}]), for iodine, at specified energies (`E`). This is illustrated in Fig. 16.3 , where cross marks indicate tabulated values and solid lines indicate interpolated values at uniform 0.01 keV energy spacings. Interpolation discontinuities exist at absorption edges in mass attenuation and energy absorption coefficients and photoelectric cross sections.

Mass coefficients for elements can be combined to describe complex materials. Many standard text books such as Johns and Cunningham[139] or Attix[140] describe the physics of this in detail. The function `xrCoefFormula` returns a vector of coefficients for any material with a specified chemical composition, such as:

```
% Interpolation of x-ray interaction coefficients based on chemical formula

E = 1:.01:50;
formula = 'Gd2O2S';
mu = xrCoefFormula(E, formula, 'attenuation');
```

The default output of `xrCoef` are mass attenuation coefficients, but other optional inputs can be specified to return (energy) "absorption" coefficients [cm^2/g] and three atomic interaction cross-sections, "sigmaPhotoelectric", "sigmaCoherent" and "sigmaIncoherent" [barns/atom]. For convenience, the three cross sections are also provided as mass coefficients after scaling by the atomic weight (atoms/g) for each element.

Transmission of a spectrum $q_o(E)$, through a material having uniform attenuation coefficient $\frac{\mu}{\rho}(E)$, mass density ρ and thickness x, results in the transmitted spectrum $q(E)$, where

$$q(E) = q_o(E)e^{-\frac{\mu}{\rho}(E)\rho x}. \tag{16.3}$$

This is accomplished programmatically using the xrTk function **xrTrans**. The mass density of pure elements at standard temperature and pressure can be obtained using **xrMaterialInfo**. This is illustrated in the segment below:

```
% Transmission of x-ray beam with spectrum q0 through an attenuator

E = 10:.25:120; kV = 90;

AlInfo = xrMaterialInfo('aluminum');  % Attenuator material data
AlThickness = 10.;  % Attenuator thickness [cm]
qi0 = xrSpec(E, kV);  % Spectrum incident on attenuator
massLoading = AlThickness*AlInfo.m_density; % Mass-loading [cm2/g]
qi = qi0.*xrTrans(E, 'aluminum', massLoading);  % Transmitted spectrum
```

When **xrTrans** is passed a complex material, transmission is calculated by element-by-element multiplication by **xrTrans** with appropriate mass loading (g/cm^2) values for each element.

16.2.3 Exposure and dose

Patient exposure to radiation is often expressed as the absorbed dose (absorbed energy per unit mass) in air at the patient entrance surface. At diagnostic energies, this is given by air KERMA, K [Gy][139]:

$$K = 1.6022 \times 10^{-19} \left[\frac{J}{eV} \right] \times 10^8 \left[\frac{eV}{keV} \frac{mm^2}{cm^2} \frac{g}{kg} \right] \int_0^{kV} q(E) \frac{\mu_{en}}{\rho}{}_{air} (E) E dE \tag{16.4}$$

where $q(E)$ is the spectrum incident on the patient and $\frac{\mu_{en}}{\rho}{}_{air}(E)$ is the mass energy-absorption coefficient for air. With xrTk, spectral intensity can be matched to a desired air KERMA and the value of Q_o [quanta mm^{-2} Gy^{-1}] calculated as below:

```
% Scale spectrum qo to 2.5 uGy air KERMA and calculate Qo (quanta/mm2 per Gy)

E = 10:.25:120; kV = 90;
qi0 = xrSpec(E, kV);

f = 2.5e-6/xrAirKerma(E, qi0); % Conversion factor to 2.5 uGy air KERMA
qi = qi0*f; % Normalized spectrum with 2.5 uGy air KERMA

K = xrAirKerma(E, qi)*1e6;
Qo = sum(qi)/K;
```

16.2.4 Image contrast

Regional differences in x-ray transmission through an attenuator (i.e., patient) cause variations in the fluence at the detector surface. The detector generally consists of either a scintillating converter layer (e.g., cesium iodide or gadolinium oxisulphide) that uses x-ray energy to liberate light quanta, coupled to a photo-sensor array ("indirect" design), or a photoconductor (e.g., selenium) that uses x-ray energy to liberate charge, coupled to a charge-sensor array ("direct" design)[141]. The sensor arrays consist of a matrix of elements with each element producing a signal value proportional to the x-ray energy deposited in the converter at the element position. For example, iodinated vessels filled with an angiographic

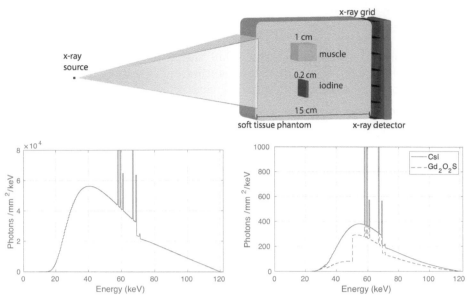

Figure 16.4: Illustration of an x-ray beam incident on 15-cm of soft-tissue containing a 1-cm thick muscle region and 0.2-cm iodinated region (0.1 g/cm³ iodine). Spectra are shown of x-rays incident on the phantom (*left*) and interacting in CsI and Gd₂O₂S-based detectors (*right*).

contrast agent in a uniform soft-tissue background (see Figure 16.4) cause image contrast C, where

$$C = \frac{d - d_b}{d_b} \tag{16.5}$$

and where d and d_b are detector signals corresponding to iodinated and background regions, respectively. Detector signal is proportional to the energy absorbed and the background signal averaged over the x-ray spectrum is,

$$d_b = kfa \int_0^{kV} q_o(E) e^{-\frac{\mu}{\rho}_b(E)\rho_b x_b} G\alpha(E) E \mathrm{d}E \tag{16.6}$$

where the background region is assumed to have a mass attenuation coefficient of $\frac{\mu}{\rho}_b$ and a density and thickness ρ_b and x_b, respectively. The factor k (pixel-value per unit energy deposited) is a constant of proportionality, f is the detector fill factor (fraction of detector area sensitive to radiation), G is the primary-beam transmission factor through the anti-scatter grid (assuming full attenuation of all scatter) and a is detector element area. The quantum efficiency of the detector α is given by,

$$\alpha(E) = 1 - e^{-\left(\frac{\mu_{pe}}{\rho}(E) + \frac{\mu_{inc}}{\rho}(E)\right)\rho_d x_d}, \tag{16.7}$$

where $\frac{\mu_{pe}}{\rho}$ and $\frac{\mu_{inc}}{\rho}$ are the photoelectric and incoherent scattering mass interaction co-efficients of the *detector converter layer*, respectively, with density ρ_d and thickness x_d. Detector signal in the iodinated region is therefore given by,

$$d = kfa \int_0^{kV} q_o(E) e^{-\left(\frac{\mu}{\rho}(E)\rho x + \frac{\mu}{\rho}_b(E)\rho_b(x_b - x)\right)} G\alpha(E) E \mathrm{d}E, \tag{16.8}$$

where x is the thickness of the region with contrast agent and $\frac{\mu}{\rho}$ is associated mass attenuation coefficient. These equations can be combined to predict the image contrast resulting from the iodinated regions. This result assumes all x-ray energy from each interacting x-ray photon is absorbed in the converter layer including absorption of characteristic emissions from photoelectric interactions and Compton scattered x-rays. Replacing the term E with $E_{ab}(E)$, which describes the average energy absorbed per incident x-ray of energy E in a specific material, would describe the signal if characteristic emissions and Compton scatter escape the converter.

The following code segment illustrates how the quantities in Eqs. (16.5) to (16.8) could be evaluated using xrTk:

```
% Image contrast from iodine in soft tissue

% Background region material, thickness [cm] and density [g/cm3]
materialb = 'Tissue_Soft_ICRU-44'; xb = 15; densityb = 1.06;

% Iodinated material, thickness [cm] and density [g/cm3]
material = 'iodine'; x = 0.2; density = 0.1;

% Generate spectrum
E = 10:.5:150;
kV = 50;
qi = xrSpec(E, kV);

% Detector converter material, thickness [cm], density [g/cm3]
materiald = 'CsI'; xd = 0.03; densityd = 4.7;
infod = xrMaterialInfoFormula(materiald);

fprintf('Detector: %s\n', materiald);
fprintf('Soft tissue: %s\n', materialb);
fprintf('Contrast material: %s\n', material);

% Pixel area [cm2], k-factor, CsI fill-factor and grid-factor
a = 0.1*0.1; k = 1;
f = 0.8; % Fill factor less than unity due to hexagonal packing of CsI
G = 0.5; % Grid factor

% Quantum efficiency
mu_pe = xrCoefFormula(E, materiald, 'photoelectric'); % [cm^2/g]
mu_inc = xrCoefFormula(E, materiald,'incoherent'); % [cm^2/g]
alpha = 1-exp(-(mu_pe+mu_inc)*densityd*xd); % Eq. 16.7

% Detector signals
db = k*f*a*sum(qi.*xrTrans(E, materialb, xb*densityb).*G.*alpha.*E); % Eq. 16.6
d = k*f*a*sum(qi.*xrTrans(E, materialb, (xb-x)^densityb) ...
    .*xrTrans(E, material, x*density).*G.*alpha.*E); % Eq. 16.8

% Contrast
C = (d-db)/db; % Eq. 16.5

fprintf('Contrast with 50-kV spectrum: %f\n', C);
```

The implementation of the integrals (summations) in Eqs. (16.5) to (16.8) are well behaved in the sense the integrands approach zero for both low and high energy limits, and each integral is evaluated by summing appropriate vector point-wise products.

Object contrast is a complicated function of x-ray spectral shape and xrTk can be used to find exposure conditions that optimize the contrast of specific materials. Figure 16.5 shows

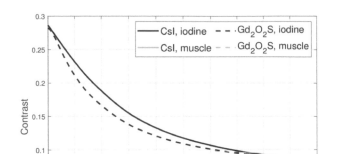

Figure 16.5: The magnitude of image contrast from iodine (0.2 cm, 0.1 g/cm³) and muscle (1 cm, ICRU Report 44) in a 15-cm soft tissue phantom (ICRU Report 44) expected with CsI and Gd_2O_2S-based detectors as a function of kV setting. As expected, the iodinated region has greater contrast than muscle and contrast decreases with increasing kV.

iodine and muscle contrast relative to soft tissue (see ICRU Report 44[142]) as a function of kV setting for CsI and Gd_2O_2S-based detectors. The iodinated region has much greater contrast than the muscle region, which explains why radiography generally has modest soft-tissue sensitivity while iodinated contrast agents used in angiography give arteries greater contrast. The contrast decreases with increasing kV for both materials due to a decrease in the relative probability of photoelectric interactions.

16.2.5 Rose signal-to-noise ratio

The detection of x-ray quanta is governed by Poisson statistics such that the variance in a number detected is equal to the mean. These random fluctuations result in image "noise" that may obscure low-contrast structures, particularly small structures. Albert Rose[143] showed this visual detection task is described by what we now call the "Rose SNR" given by[144]:

$$\text{SNR}_{\text{Rose}} = \frac{\text{mean signal}}{\text{background noise}} = C\sqrt{A\bar{q}_b} \qquad (16.9)$$

where A is the area of a uniform disk of contrast $C = (\bar{q} - \bar{q}_b)/\bar{q}_b$ and $\bar{q}_b$ is the mean number of quanta per mm² in a uniform background. Rose showed that an SNR of five is required for confident visual detection[143], which explains why small lesions are more difficult to see in a noise-limited image than large lesions of the same contrast.

The Rose SNR can be adapted to describe visual detection in a digital image. The Rose SNR for a uniform disk of N (uncorrelated, linear, energy-integrating) pixels is given by:

$$\text{SNR}_{\text{Rose}} = \frac{N(\bar{d} - \bar{d}_b)}{\sqrt{N\sigma_b^2}} = C\sqrt{N}\frac{\bar{d}_b}{\sigma_b}, \qquad (16.10)$$

where $\bar{d}$ is given by Eq. (16.8), $C = (\bar{d} - \bar{d}_b)/\bar{d}_b$, and σ_b is determined from[145]

$$\sigma_b^2 \approx k^2 f a \int_0^{\text{kV}} \bar{q}_o(E) e^{-\frac{\mu}{\rho}_b(E)\rho_b x_b} G\alpha(E)E^2 dE. \qquad (16.11)$$

This result includes Swank noise[146] resulting from the use of a polyenergetic x-ray beam

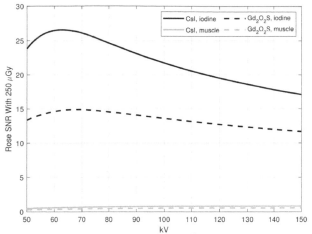

Figure 16.6: The magnitude of Rose-SNR for a 1 cm^2 region of interest of iodine and muscle from CsI and Gd$_2$O$_2$S-based detectors as a function of kV setting for constant (250 µGy) patient-entrance air-KERMA.

but does not include other sources of noise such as variations in scatter escape, absorption of scatter photons, quantum sinks, electronic readout circuits, etc. The assumption of uncorrelated pixel values is often a poor assumption, particularly with small image pixels. See Burton et al.[147] for a description of how to bin pixel values to satisfy the assumption of uncorrelated noise.

We will calculate Rose SNR (excluding the effects of scatter radiation) by generating x-ray spectra from a tungsten tube with tube potentials ranging from 50 to 150 kV and scaling the spectrum to achieve an entrance air-KERMA of 250 µGy, which is typical for an abdomen x-ray[148]. The left plot of Figure 16.4 shows a 120 kV spectrum incident on the tissue phantom. The detector signal summed over a 1 cm^2 region of interest was calculated for CsI (0.141 g/cm^2) and Gd$_2$O$_2$S (0.05 g/cm^2) detectors[149, 150]. The right-hand plot in Figure 16.4 shows the spectrum of x-rays transmitted through tissue, iodine and interacting in each converter.

Figure 16.6 shows iodine Rose SNR in a 1 cm^2 region of interest is much greater than muscle for both converters. Muscle SNR is less than 5 and therefore does not satisfy the Rose criteria for confident detection. The SNR with a CsI detector is almost twice that of Gd$_2$O$_2$S for kV values normally used for angiography. This is due to superior quantum efficiency at the energies where iodine provides the greatest contrast (above the iodine K-edge energy of 33 keV). This means that low-exposure (quantum-noise-limited) images with CsI detectors will show iodinated regions with superior SNR for the same exposure. Below is the MATLAB code and xrTk functions used for calculating iodine SNR:

```
% Iodine contrast and SNR with a CsI detector

dE = 0.5; % Energy bin spacing size [keV]
E = 10:dE:150; % Energy vector, from 10 to 150 keV with dE bin size [keV]

tissue = 'Tissue Soft_ICRU-44';
tissue_cm = 15; % Background attenuator (soft-tissue) thickness [cm]
tissue_density = 1.06; % Soft-tissue density [g/cm^3] from ICRU-44
```

(code continues on next page)

```matlab
iodine = 'iodine';
iodine_cm = 0.2; % Thickness [cm] of iodine
iodine_density = 0.1; % Density [g/cm^3] of iodine

% X-ray coefficients and quantum efficiency of detector
materiald = 'CsI';
infod = xrMaterialInfoFormula(materiald); % Material info of cesium iodide
massLoading = 0.141; % Detector mass loading [g/cm^2]
mu_pe = xrCoefFormula(E, materiald, 'photoelectric'); % [cm^2/g]
mu_inc = xrCoefFormula(E, materiald,'incoherent'); % [cm^2/g]
alpha = 1-exp(-(mu_pe+mu_inc)*massLoading); % Eq. 7

kVSet = 50:1:150; % Range of kV settings used to generate different spectra

contrast = zeros(1, length(kVSet)); % Vector for contrast values for each kV
snr = zeros(1, length(kVSet)); % Vector for SNR values for each kV

k = 1; % Arbitrary detector element gain, [0]
fCsI = 0.8; % Detector CsI fill (packing) factor, [0]
a = 0.01*0.01;   % Element size is 0.1 mm, element area [cm^2]
N = 100*100; % Number of pixels, 100x100 pixels, ROI area = 1 cm^2
G = 0.5; % Anti-scatter grid factor [0]
M = 115/100; % Geometric magnification from phantom entrance to detector

i=1; % Counter and index for iteration through kV set
fprintf('Detector=%s\n', materiald);
for kV=kVSet
    qi0 = xrSpec(E, kV, 'W'); % Beam from W tube 1 m from source [photons/mm2/keV]
    qin = 250e-6*qi0/xrAirKerma(E, qi0); % Normalized spectrum to 250 uGy air-KERMA

    % Quanta per mm2 transmitted through tissue and incident on detector
    qib = G*(M^-2)*qin.*xrTrans(E, tissue, tissue_cm*tissue_density);

    % Quanta per mm2 transmitted through tissue and iodine and incident on detector
    qi = G*(M^-2)*qin.*xrTrans(E, tissue, (tissue_cm-iodine_cm)*tissue_density) ...
        .*xrTrans(E, iodine, iodine_cm*iodine_density);

    % Detector signal, Eqs. 6 and 8
    db = k*fCsI*a*sum(qib.*alpha.*E);
    d = k*fCsI*a*sum(qi.*alpha.*E);

    % Variance, Eq. 11
    varb = (k^2)*fCsI*a*sum(qib.*alpha.*E.*E);
    var = (k^2)*fCsI*a*sum(qi.*alpha.*E.*E);

    % Contrast as a function of x-ray beam kV, Eq. 5, made positive
    contrast(i) = (db-d)/db;

    % SNR as a function of x-ray beam kV, Eq. 10
    snr(i) = contrast(i)*sqrt(N)*db/sqrt(varb);

    fprintf('kV=%d contrast=%f SNR=%f\n', kV, contrast(i), snr(i));

    i = i+1; % Increment counter
end
```

16.3 DISCUSSION

The calculations for contrast and SNR presented here ignore the effects of scatter radiation and therefore may be accurate only when using a large air gap or very good scatter-rejection grid (with an increase in patient exposure due to grid transmission as indicated in Eq. (16.6)). Scatter will add image noise due to the detection of scatter photons and also reduce image contrast. Approximate results can be obtained by adding a scatter component to the fluence spectrum $q(E)$, that is approximately uniform throughout regions of the image. Regardless, results obtained using this simple toolkit provide a good first estimate of image contrast, signal and noise, and provide important insight to optimization problems. MATLAB provides a programming environment that is well suited to these vector-based calculations.

The x-ray coefficients and material data used by xrTk come from Hubbell and the NIST group, which have an estimated uncertainty of 2%[151]. X-ray spectra are generated using the Tucker-Barnes algorithms[135, 136] for both custom spectra with arbitrary kV settings and standard IEC spectra that have matched HVL [mmAl] and Q_o [quanta/mm^2/µGy] values that agree with IEC values within 1% for radiography and 3% for mammography. This makes xrTk the first toolkit that generates both arbitrary-kV and standardized IEC spectra having arbitrary energy-bin size for both radiography and mammography, with open-source MATLAB code.

16.4 CONCLUSIONS

In this chapter, we demonstrated how image iodine contrast and SNR and air-KERMA can be calculated with xrTk and be used to determine optimal exposure techniques in angiography. It was shown that CsI-based detectors provide substantially higher iodine SNR than Gd$_2$O$_2$S-based detectors for the same patient air KERMA. The optimal x-ray energy with an average-thickness adult is approximately 60 kV with little decrease in SNR per unit air KERMA up to 80 kV. Results may be different for other patient sizes or if scatter is included in the calculation.

MATLAB toolboxes used in this chapter:
None

Index of the in-built MATLAB functions used:

exp	length	sum	zeros
fprintf	sqrt		

## CHAPTER 17

Automating daily QC for an MRI scanner

Sven Månsson

Medical Radiation Physics, Department of Translational Medicine, Lund University, Skåne University Hospital, Malmö, Sweden

CONTENTS

THIS chapter is intended for medical physicists or engineers who want to automate the analysis of images acquired during quality control. The example will show how images of a dedicated quality control phantom (the ACR MRI accreditation phantom) can be used to examine slice thickness, image uniformity and geometric distortion.

17.1 INTRODUCTION

When performing quality control on an MR scanner, a variety of tests are available. This chapter will demonstrate tests for slice thickness, signal uniformity and geometric distortion. The geometric distortion is mainly a test of the gradient system, the slice thickness tests both the gradient and the RF systems, while the uniformity can be seen as a global system test[152]. Other, more time-consuming tests are measurements of homogeneity and drift of the main magnetic field, measurement of RF transmitter gain and stability, and MR spectroscopy tests.

In the following sections, MATLAB code will be presented which performs the image analysis automatically. For simplicity, user input is needed to locate the image directory, but this may also be automated in a real-world procedure. The image analysis relies on a standardized imaging protocol, on a standardized phantom and on a standardized routine for placing the image planes. Using this protocol, the images for quality control can be collected in as little as ten minutes, including the time for placing and removing the phantom and the RF-coil.

17.1.1 The ACR MRI accreditation phantom

The ACR MRI accreditation phantom is a cylindrical phantom (diameter 20.4 cm, length 16.5 cm) designed to fit inside a head RF coil in axial, sagittal and coronal orientations, thereby allowing tests in all three major planes. The phantom contains sub-sections allowing evaluation of slice position, uniformity, geometrical distortion among others. More details about the ACR MRI accreditation phantom and the possible quality control tests can be found in the overview by G.D. Clarke[153] or the Phantom Test Guidance from the American College of Radiology (ACR)[154].

17.1.2 Imaging protocol

The minimum imaging protocol for the ACR MRI accreditation phantom consists of a localizer scan and one multi-slice scan for quality control. The ACR test guidance, however, prescribes two multi-slice scans for quality control: one T1-weighted and one T2-weighted. In this chapter it is assumed that the ACR phantom is oriented axially, that is, the multi-slice scan is also axial and the localizer scan is sagittal. Depending on the ambition of the quality control and the time available, several types of multi-slice images (spin echo, turbo spin echo, gradient echo, T1-weighted, T2-weighted multi-echo, etc.) may be acquired, and the imaging may be repeated in the sagittal and coronal orientations. To simplify the automatic image analysis, it relies on the multi-slice images being acquired with a standardized position for the first slice, and a mandatory slice thickness of 5 mm with 10 mm inter-slice distance (5 mm gap). In this way, it is assured that a certain slice can be used for a certain test, for example, slice thickness test in slice #1 and signal uniformity test in slice #7. The images shown in this chapter were acquired with the following parameters: TR/TE = 500/10 ms, 12 slices, 256 × 256 matrix, FOV = 25 × 25 cm^2, 1 average, scan time 2 minutes. Note that any filters or post-processing steps (smoothing, sharpening, zero-filling) on the scanner should be turned off.

17.2 AUTOMATIC ANALYSIS OF QUALITY CONTROL IMAGES

In the remainder of the chapter, MATLAB code (using the *Image Processing Toolbox*) will be outlined for automatic evaluation of slice thickness, signal uniformity and geometric distortion of the ACR MRI accreditation phantom. The reader may extend the code for example, high-contrast spatial resolution test (slice #1), ghosting ratio measurement (slice #7), or low-contrast object detectability (slices #8-11). The only manual interaction needed is the selection of the directory in which the multi-slice data set is located. To obtain correct results, images must be acquired with correct slice position along the z-direction according to the manual for the ACR MRI accreditation phantom. However, the positioning of the phantom in the x- and y-directions and its rotation angle are not critical, as the script will detect and correct for deviations from the optimal position. The total execution time for the three tests was 2–3 s on a standard laptop PC. The images were acquired on a Siemens scanner, which stores each slice as an individual DICOM file. To load images from a different scanner vendor, modifications of the script may be necessary.

17.2.1 Slice thickness test

If the positioning of the multi-slice scan was done properly, slice #1 should be similar to Figure 17.1a. However, the automatic analysis must be able to handle also images where the phantom has been placed less perfectly, so for this example the image in Figure 17.1b will be used.

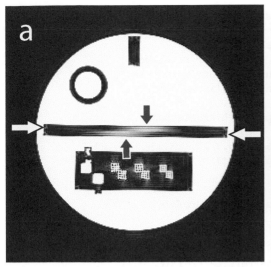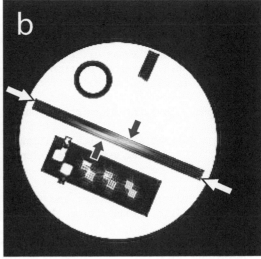

Figure 17.1: Slice #1 is used for the slice thickness test. The brightness has been increased for better visualization of the slits (black arrows). a) The goal is to measure the intensity profile along the two horizontal bars (white arrows). b) For the example, an image will be used where the phantom has been shifted and rotated roughly 10° clockwise.

Two bars (indicated by white arrows in Figure 17.1a), each with a narrow slit, appear in slice #1. The slits have a slope of 10:1 and will appear in the image with a length that is 10 times longer than the actual slice thickness (black arrows in Figure 17.1b). The two slits are crossed, such that averaging the measured lengths will compensate for a slight tilting of the phantom[154]. The automatic image analysis in the listing below consists of three steps:

1. Identify the bars and rotate the image such that the bars are horizontal.

2. Create a profile through each bar and measure its length in mm.

3. Calculate the measured slice thickness according to Eq. 17.1 [153]:

$$\text{sliceThickness} = \frac{1}{10} \frac{\text{Length}_{upper} + \text{Length}_{lower}}{2} \qquad (17.1)$$

These steps are performed in the function **sliceThickTest**. The result is compared against the ACR quality criterion, in which the thickness error should preferably be <0.7 mm, and an error >1.0 mm is considered a failure of the test[154]:

```
function result = sliceThickTest(imagePath)
%% Load slice #1
info = dicominfo(imagePath{1});
data = double(dicomread(info));
result.dateAndTime = [info.SeriesDate, ' ', info.SeriesTime(1:6)];
% Scale image between 0.0 and 1.0
dataScaled = data/max(data(:));
result.originalImage = dataScaled;
```

(code continues on next page)

```matlab
% Use the Hough transform to detect the angle of the thickness bars
% and create a rotation-corrected image.
% "verticalPosition" are the coarse y-positions of the
% upper and lower edges of the thickness bars (after rotation)
[rotationAngle, verticalPosition] = doHoughTransform(dataScaled);
dataRot = imrotate(dataScaled, rotationAngle, 'bilinear','crop');

% Cut out a horizontal stripe surrounding the thickness bars
margin = 6;
stripe = dataRot(min(verticalPosition)-margin: ...
                 max(verticalPosition)+margin, :);
result.rampImage = stripe;

% Average the rows of the stripe to get a vertical profile
verticalProfile = mean(stripe, 2);

% Cut away the bright areas above and below the dark bars
lowIntensityIndices = findBackgroundPoints(verticalProfile, 0);
stripe = stripe(lowIntensityIndices, :);

% Separate the upper from the lower bar and average the columns of each,
% with some upper and lower margins
height = size(stripe,1);
upperStripe = mean(stripe(2:floor(height/2)-1, :), 1);
lowerStripe = mean(stripe(ceil(height/2)+1:height-2, :), 1);

% Analyze each profile for slice thickness
dx = info.PixelSpacing(2);
upperThickness = calculateSliceThickness(upperStripe, dx);
lowerThickness = calculateSliceThickness(lowerStripe, dx);
result.sliceThickness = mean([upperThickness lowerThickness]);
result.barThicknesses = [upperThickness lowerThickness];

% Check if the test passed the ACR's tolerance
ST = result.sliceThickness;
if abs(ST - 5) > 1
    fprintf('TEST FAILED. Deviation > 1.0 mm: %1.2f mm.\n', ST)
elseif abs(ST - 5) > 0.7
    fprintf('WARNING, deviation > 0.7 mm: %1.2f mm.\n', ST)
else
    fprintf('Slice thickness test PASSED\n')
end
end % sliceThickTest
```

In the first step, the Hough transform is used to locate straight lines. The function **doHoughTransform** converts the image to an edge image and finds the two largest peaks in the Hough transform, corresponding to the two longest edges, which in this image are the upper and lower edges of the bars. The outputs of the function are the angle (in degrees) and the vertical coordinates of the edges (after they have been rotated to a horizontal position). Then, the image is actually rotated to make the ramps horizontal:

```
function [rotationAngle, verticalPosition] = doHoughTransform(data)
binaryImage = imbinarize(data);
edgeImage = edge(binaryImage,'canny');
theta = -90:0.1:89.9; % Use angular resolution of 0.1 degrees

% Straight lines in edgeImage will produce large values in H
[H,~,rho] = hough(edgeImage,'Theta',theta);

% Search for the two largest peaks in H
P = houghpeaks(H,2);
rotationAngle = mean(theta(P(:,2))) + 90; % Input angle to imrotate

% Linear algebra to calculate the vertical position of the lines
% AFTER the image has been rotated such that the lines are horizontal
alpha = -mean(theta(P(:,2)))*pi/180; % Assume the lines are parallel
unitVector = [cos(alpha);sin(alpha)]; % Unit vector perpendicular to the lines
pointLine = unitVector*rho(P(:,1)); % Calculate a point on each line
centerPoint = [size(data,1);-size(data,2)]/2 + 0.5; % The center point of the image
vector = pointLine - [centerPoint centerPoint]; % Vector from centerPoint...
                                                 % to pointLine
distance = dot([unitVector unitVector],vector); % Distance from centerPoint
                                                 % to lines
% y-indices of the lines after rotation
verticalPosition = -round((centerPoint(2)+distance));
end % doHoughTransform
```

In the second step, the vertical positions from **doHoughTransform** are used to coarsely select the thickness bars from the rest of the image (the 2D array: **stripe**). The rows of this stripe are averaged to create a vertical profile (the 1D array: **verticalProfile**) and the function **findBackgroundPoints** is used to find and remove the bright background areas above and below the bars (the **stripe** array is overwritten). Next, the updated **stripe** is split into two regions (see Figure 17.2a) and the image columns inside the bars are averaged to create a horizontal intensity profile through each bar (**upperStripe** or **lowerStripe**). Figure 17.2b illustrates such a profile.

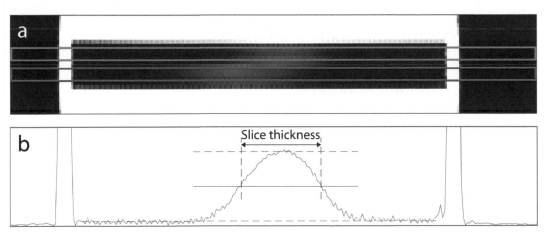

Figure 17.2: (a) The upper (*blue*) and lower (*red*) ramps which are used to create intensity profiles. (b) The slice thickness is defined as the full width at half maximum (FWHM) of a horizontal profile.

```matlab
function index = findBackgroundPoints(signal, skipExtra)
dsignal = diff(signal);
midpoint = round(length(dsignal)/2);

% Find the position of sharp transitions between dark and bright
% when moving from left to right along the signal profile
[~,startIndex] = min(dsignal(1:midpoint));  % Bright to dark
[~,stopIndex] = max(dsignal(midpoint:end)); % Dark to bright
startIndex = startIndex + 1 + skipExtra;
stopIndex = stopIndex + midpoint - 1 - skipExtra;
signal = signal(startIndex:stopIndex);
background = signal;
bNdxOld = [];
maxIter = 50; % Prevent endless loop

for iter = 1:maxIter
   backgroundLevel = median(background);
   standardDev = std(background);
   threshold = min(backgroundLevel + 2.5*standardDev,4*backgroundLevel);
   bNdx = find(signal < threshold);
   background = signal(bNdx);
   if isequal(bNdx, bNdxOld), break, end
   bNdxOld = bNdx;
end

if iter == maxIter
   warning('findBackgroundPixels: maximum number of iterations reached.')
end

% Return the indices of the background pixels along the signal profile
index = find(signal < threshold) + startIndex - 1;
end % findBackgroundPoints
```

The function `calculateSliceThickness` finds the full width at half maximum (FWHM) of the profile to calculate the slice thickness according to Eq. 17.1:

```matlab
function thickness = calculateSliceThickness(signal, dx)
% Skip 5 points at both ends of the signal profile
% and calculate the background level
backgroundIndices = findBackgroundPoints(signal, 5);
backgroundLevel = median(signal(backgroundIndices));

signal = signal(backgroundIndices(1):backgroundIndices(end));
[~,topNdx] = max(signal);
topLevel = mean(signal(topNdx-1:topNdx+1));
midLevel = mean([backgroundLevel topLevel]);
midNdx = find(signal>=midLevel);

% Calculate width in mm
FWHM = length(midNdx(1):midNdx(end))*dx;
thickness = FWHM/10;
end % calculateSliceThickness
```

17.2.2 Uniformity test

Slice #7 has no structures inside, except for a small rectangle near the circumference (Figure 17.3). The function `uniformityTest` first applies a 3×3 median filter to reduce the noise sensitivity of the measurement. Next, it uses the function `placeCircularROI` to place a circular ROI which covers 65% of the phantom area. If the signal intensity within the ROI is denoted S, the image intensity uniformity (in percent) is calculated as:

$$Uniformity = 100 \left(1 - \frac{\max(S) - \min(S)}{\max(S) + \min(S)} \right) \tag{17.2}$$

The result is compared against the ACR quality criterion, in which the uniformity should preferably be >87.5%, and the test fails if the unifomity is <85% for a scanner with field strength below 3T[154]. The corresponding limits for a 3T scanner are 82% and 80%.

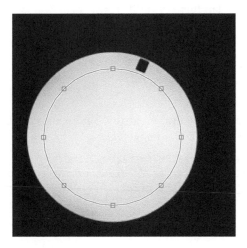

Figure 17.3: Slice #7 is used for the uniformity test.

```
function result = uniformityTest(imagePath)
%% Load slice #7
dcmInfo = dicominfo(imagePath{7});
dcmData = double(dicomread(dcmInfo));
result.originalImage = dcmData;

% Apply a 3x3 median filter to reduce the noise sensitivity
dcmDataFiltered = medfilt2(dcmData);

% Show the image
figure(1), clf, colormap(gray(255)), ax = axes;
imagesc(ax, dcmDataFiltered)
set(gca, 'DataAspectRatio', [1 1 1])

% Create a circular ROI. The ROI covers 65% of the phantom area
pos = placeCircularROI(dcmDataFiltered, sqrt(0.65));
result.ROIpos = pos;
ROI = imellipse(ax, pos);
ROImask = ROI.createMask; % createMask is a MATLAB built-in function
```

(code continues on next page)

```
% Get the highest and lowest signal intensities from the ROI
maxSignal = max(dcmDataFiltered(ROImask));
minSignal = min(dcmDataFiltered(ROImask));
result.uniformity = 100*(1 - (maxSignal - minSignal)/(maxSignal + minSignal));
result.signalIntensities = [maxSignal minSignal];

% Check if the test passed the ACR's tolerance
IIU = result.uniformity; % Image intensity uniformity
if dcmInfo.MagneticFieldStrength < 3
   tol = [85 87.5];
else
   tol = [80 82];
end
if IIU < tol(1)
   fprintf('TEST FAILED. Uniformity < %1.1f%%: %1.1f%%.\n', tol(1), IIU)
elseif IIU < tol(2)
   fprintf('WARNING, uniformity < %1.1f%%: %1.1f%%.\n', tol(2), IIU)
else
   fprintf('Uniformity test PASSED\n')
end
end % uniformityTest
```

The function `placeCircularROI` first creates a binary mask using a threshold at the mean signal level of the bright parts of the image:

```
function [pos,r0] = placeCircularROI(data,relDiameter)
% relDiameter is the ROI diameter as fraction of the phantom diameter
% compute a mask which finds bright pixels
mask = findBrightPoints(data);

% Find the extent of the mask
px = find(any(mask,1));
py = find(any(mask,2));

% Find the length and center of the mask. All lengths are in pixel units
xCenter = mean([px(1) px(end)]);
xLength = px(end) - px(1) + 1;
yCenter = mean([py(1) py(end)]);
yLength = py(end) - py(1) + 1;

phantomDiameter = mean([xLength yLength]);
ROIdiameter = phantomDiameter*relDiameter;
xMin = xCenter - ROIdiameter/2;
yMin = yCenter - ROIdiameter/2;
pos = [xMin yMin ROIdiameter ROIdiameter];
r0 = [xCenter yCenter];
end % placeCircularROI
```

The threshold and mask are calculated by the function `findBrightPoints`:

```
function mask = findBrightPoints(data)
% Compute a mask which finds bright pixels
Hi = mean(data(:));
Lo = 0;
oldThreshold = 0;
```

(code continues on next page)

```
for k = 1:20 % Prevent endless loop
    threshold = (Hi+Lo)/2;
    mask = data > threshold;
    d = abs(threshold - oldThreshold)/threshold;
    if d < 1e-4, break, end
    oldThreshold = threshold;
    Hi = mean(data(mask));
    Lo = mean(data(~mask));
end
end % findBrightPoints
```

17.2.3 Geometric distortion test

The geometric distortion test compares measured distances against ground truth. In this test, the diameter of the phantom is mesured in four directions: horizontally, vertically and along the diagonals. Slice #5 was used (Figure 17.4). This slice contains a square grid pattern to facilitate manual distance measurements, but the automatic measurement can in principle be made on any slice.

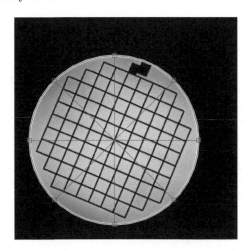

Figure 17.4: Slice #5 is used for the geometric distortion test. The diameter is measured along the four red lines.

To find the center and approximate diameter of the phantom, the function **geoDistTest** first calls **placeCircularROI**. Then it places four radial profiles extending outside the phantom and uses the MATLAB function **interp2** to find the intensity values along the profiles. To increase the accuracy of the mesurement, the profiles are interpolated to a resolution four times higher than the pixel resolution.

```
function result = geoDistTest(imagePath)
%% Load slice #5
info = dicominfo(imagePath{5});
data = double(dicomread(info));
result.originalImage = data;
```

(code continues on next page)

```matlab
% Show the image
figure(2), clf, colormap(gray(255)), ax = axes;
imagesc(ax, data)
set(gca,'DataAspectRatio',[1 1 1])

% This routine only works for isotropic image resolution
assertMsg = 'geoDistTest: cannot handle non-isotropic resolution';
assert(info.PixelSpacing(1)==info.PixelSpacing(2), assertMsg)
pixelSpacing = mean(info.PixelSpacing);

[pos, center] = placeCircularROI(data ,1);
ROI_c = imellipse(ax, pos);

% Draw 4 lines: vertical, horizontal and diagonals
alpha = [90 -45 0 45]*pi/180;
directions = {'vertical', 'NE-SW diagonal', 'horizontal', 'NW-SE diagonal'};
D = 0.25; % Profile resolution 0.25 pixel
radius = pos(3)/2;
r0 = 0:D:1.15*radius; % Define a radial spoke

for k = 1:length(alpha)
    % x- and y-coordinates for the lines expressed in complex notation: r = x + j*y
    r = r0*exp(j*alpha(k));
    r = r(2:end);
    % Extend the spoke to cover the full diameter of the phantom
    r = [fliplr(-r) 0 r] + center(1) + j*center(2);

    % Get the intensity profile along the spoke
    profile = interp2(data, real(r), imag(r));

    % Find the length of the profile and convert to mm
    [lengthInPixels, mask] = findProfileLength(profile, D);
    profileLength(k) = lengthInPixels*pixelSpacing;

    % Draw the lines
    line(ax, real(r(mask)), imag(r(mask)), 'color', 'r');
    result.r{k} = r(mask);
end

% The true diameter of the ACR phantom is 190 mm
trueProfileLength = 190;
deviation = profileLength - trueProfileLength;
[~, ndx] = max(abs(deviation));
result.maxDeviation = deviation(ndx);
result.direction = directions{ndx};
result.profileLength = profileLength;
result.trueProfileLength = trueProfileLength;

% Check if the test passed the ACR's tolerance
md = abs(result.maxDeviation); % maximum deviation from true length
if md > 3
```

(code continues on next page)

```
    fprintf('TEST FAILED. Maximum deviation > 3 mm: %1.1f mm.\n', md)
elseif md > 2
    fprintf('WARNING, maximum deviation > 2 mm: %1.1f mm.\n', md)
else
    fprintf('Geometric distortion test PASSED\n')
end
end % geoDistTest
```

The function `findProfileLength` calls the function `findBrightPoints` to find the begining and end of each profile and calculates its length. The ACR quality criterion states that all measured lengths should preferably be within ±2 mm of their true values, and the test fails if any deviation is larger than 3 mm[154].

```
function [L, mask] = findProfileLength(signal, D)
mask = findBrightPoints(signal);
L = find(mask);
L = (L(end) - L(1))*D;
end % findProfileLength
```

17.3 THE MAIN FUNCTION

To wrap up the quality controls described in the previous sections, the function `automaticQualityControl` loads the multi-slice data set and sorts the images according to their slice location, before calling the previously described quality control test functions:

```
function result = automaticQualityControl
%% Locate the directory where the multi-slice data set is stored
defaultDirectory = 'ExampleDataChapter17';
imageDirectory = uigetdir(defaultDirectory, 'Locate multi-slice dataset');
if imageDirectory==0, return, end

% Store the path to each DICOM image
directoryList = dir(imageDirectory);
if isempty(directoryList), return, end
imageNum = 0;
for fileNum = 1:length(directoryList)
    if strcmp(directoryList(fileNum).name, '.') || ...
            strcmp(directoryList(fileNum).name, '..')
        continue
    end
    fileName = [imageDirectory filesep directoryList(fileNum).name];
    if isdicom(fileName)
        imageNum = imageNum+1;
        qualityControlImagesPath{imageNum} = fileName;
        % Find and store the slice location
        info = dicominfo(fileName);
        if isfield(info,'SliceLocation')
            sliceLocation(imageNum) = info.SliceLocation;
        else
            error('Could not detect slice location of file %s',fileName)
        end
    end
end
```

(code continues on next page)

```
% Sort image paths according to slice location, starting with lowest value
[~,index] = sort(sliceLocation, 'ascend');
qualityControlImagesPath = qualityControlImagesPath(index);

% Do quality control tests
result.sliceThicknessTest      = sliceThickTest(qualityControlImagesPath);
result.uniformityTest          = uniformityTest(qualityControlImagesPath);
result.geometricDistortionTest = geoDistTest(qualityControlImagesPath);
disp('=== automaticQualityControl completed ok ===')


end % automaticQualityControl
```

The result of the quality control, which is stored in the structure `result`, could simply be printed in the MATLAB command window, or—in a more ambitious approach—be stored in an Excel spreadsheet or in a database file. Another option would be to use the *MATLAB Report Generator* to save the results to a report in pdf or html format. Please see Chapter 5 (*Creating automated workflows using MATLAB*) for information on how to set up a database and how to use the MATLAB Report Generator.

17.4 CONCLUSION

In this chapter, MATLAB has been used to automate some image analysis tasks that can be a part of a daily QC procedure for MRI, such as testing for slice thickness accuracy, signal uniformity and geometric distortion. There are many similarities between this chapter and Chapter 11 (*Automating quality control tests and evaluating ATCM in computed tomography*). For instance, the structure of the code presented in this chapter mirrors the design philosophy of building the procedure up from a main function, which makes calls to other functions specified for certain image analysis tasks. This is beneficial for many reasons, paricularly in terms of getting a simpler overview, and the modular design allows re-use of image analysis functions in other projects.

MATLAB toolboxes used in this chapter:
Image Processing Toolbox

Index of the in-built MATLAB functions used:

abs	edge	imellipse	real
any	end	imrotate	return
assert	exp	interp2	round
break	find	isdicom	sin
ceil	fliplr	isequal	size
cos	floor	isfield	sort
dicominfo	fprintf	length	sqrt
dicomread	fullfile	line	std
diff	hough	max	strcmp
dir	houghpeaks	mean	uigetdir
disp	imag	median	
dot	imagesc	min	
double	imbinarize	pi	

## CHAPTER 18

Image processing at scale by containerizing MATLAB

James d'Arcy, Simon J Doran and Matthew Orton

Division of Radiotherapy and Imaging, Institute of Cancer Research, London, UK

CONTENTS

T HIS chapter illustrates how MATLAB can be coupled with other technologies to provide a simple, yet powerful workflow for processing data at scale. Our chosen exemplars use MATLAB's facility for calling external Java code and so-called containerization via Docker.

18.1   INTRODUCTION

Artificial Intelligence may be defined as the ability of computers or other devices to undertake activities normally associated with human cognition, for example, pattern recognition, problem solving and learning. A general increase in computing performance, coupled with recent developments in associated technologies such as convolutional neural networks [155] means that, at the time of writing, we are on the cusp of a revolution in healthcare, as machines start to take on roles that have previously required highly trained doctors. An aging population means that demand will continue to increase faster than new human practitioners can be trained and so, to address the current capacity problem in health services worldwide, it will be necessary to process many thousands of patient scans with novel algorithms.

We thus need to interact with large numbers of files in the DICOM image format. Below, we show how to navigate the relationships between data in separate files and how to locate the files containing specific images within a large data store. By encapsulating processing in a *container*, we show how MATLAB code can be launched directly from a third-party image archive platform in order to automate the processing of data from entire patient cohorts.

18.2   IMPROVED DICOM SUPPORT BY MATLAB-JAVA INTEGRATION

18.2.1   Extending the MATLAB native support for DICOM

As discussed in Chapter 3 (*Data sources in medical imaging*), the DICOM standard defines a model of the real world using concepts humans are familiar with: patients, studies, series, images, etc. Users of GUI-based DICOM software, such as a hospital Picture Archiving and Communication System (PACS), interact with these entities, whilst the files themselves are handled transparently to the user. By contrast, MATLAB's native support for DICOM is oriented towards processing named files and does not include concepts from the real-world model. A common problem facing a MATLAB programmer is thus to determine which files they need and to locate them when faced with an undocumented or variable directory structure containing thousands of DICOM entries, whose opaque file names and ordering might be unrelated to the image contents.

One of the advantages of DICOM for data storage is the richness of its metadata. Part of those metadata are UIDs (Unique IDentifiers) that are used to identify studies, series and SOP instances and define the relationships between them. An obvious method to construct the heirarchy defined in the DICOM standard would be to read every DICOM file in a directory, recursing into sub-directories if necessary, recording patient information, study/series/SOP instance UIDs and returning a set of nested MATLAB stuctures. While this is perfectly feasible in native MATLAB code, the performance is too slow for practical use on large data sets. Solving this problem was the genesis of *EtherJ*, a Java library to bring aspects of the DICOM real-world model to MATLAB in a performant and easy-to-use form.

18.2.2   The EtherJ library

EtherJ was designed by one of the authors (JD) to provide the patient/study/series/SOP instance hierarchy in a MATLAB-friendly fashion. It encapsulates the powerful DCM4CHE2 Java library, which gives high DICOM read- and write-performance[1]. Many of the interfaces in EtherJ extend from `icr.etherj.Displayable`, which provides the `display()` method to allow MATLAB users to show the contents of Java objects on the console as if they were

---

[1]Find DCM4CHE2 here: `https://www.dcm4che.org/`.

native MATLAB, leading to simpler debugging. An introduction to using Java in MATLAB is provided in an earlier chapter (*Chapter 6: Integration with other programming languages and environments*). The reader can download EtherJ from a public repository[2] and compile the Java libraries from source[3]. However, the simplest way to get started is to go to the repository for this book[4], where the pre-compiled Java archive files are provided. These can then be used in MATLAB by adding them to the static java path through adding entries in your "javaclasspath.txt" file, or, alternatively, by using the `javaaddpath` command to add them to the dynamic Java path.

With EtherJ installed, the problem of parsing a directory of DICOM files becomes a simple function such as that shown below, with the "heavy lifting" done in Java[5]. It should be noted that as `jPatientList` is an instance of `java.util.List`, the index of the first element is zero and not one as for MATLAB arrays. The authors find it convenient to prefix Java objects with "j" to clearly distinguish them from native MATLAB types.

```
function jPatientList = scanPath(path)

import icr.etherj.dicom.*;

jScanner = DicomToolkit.getToolkit().createPathScan();
jReceiver = DicomReceiver();
jScanner.addContext(jReceiver);
searchSubfolders = false;
jScanner.scan(path, searchSubfolders);
jPatientList = jReceiver.getPatientRoot().getPatientList();

end
```

Patients in the list can be queried to find studies by their study instance UID and analagous queries are possible using series instance UIDs and SOP instance UIDs, respectively. This facility to locate studies, series and SOP instances by their UID is crucial when processing other DICOM types such as RT Structure Sets or Segmentations which use UIDs to identify which DICOM entities they refer to.

The **Patient**, **Study**, **Series** and **SopInstance** interfaces provided by EtherJ's `icr.etherj.dicom` package are lightweight objects principally to establish the hierarchy of the DICOM data set and to provide a method of locating the file that contains a given SOP instance. While the contents of the file can be stored in the **SopInstance**, this is not done on directory parsing to conserve memory when parsing large data sets and can be loaded, transparently to the user, on demand. Furthermore the file contents are held by the **SopInstance** using a `java.lang.ref.WeakReference` to allow garbage collection if the Java heap memory becomes exhausted. On-demand reloading is, again, transparent to any client code.

18.2.3 A note on DICOM information object definitions and modules

The contents of a SOP instance in DICOM are determined by its SOP class and associated IOD (Information Object Definition). Each IOD is defined as a set of modules that may be mandatory, conditional or optional. The modules, in turn, are defined in terms of DICOM

---

[2]Find EtherJ here: `https://bitbucket.org/icrimaginginformatics/etherj-core`.

[3]Scripts to compile using the *gradle* build tool (see: `https://gradle.org/`) are provided in the EtherJ repository.

[4]Find the book repository here: `https://bitbucket.org/DRPWM/code`.

[5]The scanning of a path for DICOM files in EtherJ utilizes the *Observer pattern* with `jReceiver` as the observer and `jScanner` as the observed. See: `https://en.wikipedia.org/wiki/Observer_pattern` (accessed 27/11/2018).

elements that are Type 1 (required and non-empty), Type 2 (required but may be empty) and Type 3 (optional). Required elements may also be conditional and are denoted Type 1C and Type 2C. In each case, module or element, the condition is specified in the IOD or module definition. For example, in the RT Structure Set IOD the SOP Common module is mandatory and the SOP Class UID and SOP Instance UID are Type 1 mandatory elements within the SOP Common module.

EtherJ seeks to simplify the correct usage of a subset of IODs and their constituent modules by providing implementations that conform to the standard in terms of required behaviour. The `icr.etherj.dicom.iod` and `icr.etherj.dicom.iod.module` packages contain all the supported types and support for reading and writing them via a set of classes implementing the interface `org.dcm4che2.data.DicomObject`.

18.2.4 Worked examples

These worked examples demonstrate a number of usages of EtherJ Java objects and native MATLAB data types to interact with DICOM data, including the retrieval of SOP Instances and series via their UIDs and the loading of image co-ordinate data.

Retrieving DICOM objects via their UIDs

Having scanned a path for DICOM files using the `scanPath` function, it may be necessary to locate individual SOP instances in the Java object returned by `scanPath`. Identifying SOP instances is naturally done by UID and *linkages* between different SOP instances are an important feature of the way DICOM works (e.g., an individual image slice with the contours drawn on it). So it is convenient to parse the output of `scanPath` to enable direct retrieval using the UIDs. The following code is a function that returns a *map* (the first output) whose *keys* are SOP instance UIDs and whose *values* are SOP instances stored as `icr.etherj.dicom.SopInstance` Java objects[6]. The second output of this function is another map that enables the analogous retrieval of DICOM series using the Series Instance UID as a key. This function only includes MR image data in the output maps by checking the SOP class UID of each SOP instance. Other SOP class UIDs can be found from online resources such as NEMA's DICOM pages[7].

```
% Make maps for SOP instances and series that use the SOP instance UID
% and series instance UID as keys.

function [sopInstanceMap,seriesMap] = makeUIDmaps(jPatientList)

import icr.etherj.dicom.*

% Only include instances of MR image data by checking SOP class UID.
mrSopClassUid = '1.2.840.10008.5.1.4.1.1.4';
```

(code continues on next page)

---

[6]In MATLAB, a *map* is an object like a dictionary, with named *keys* that index to *values*.
[7]See: `http://dicom.nema.org/Dicom/2013/output/chtml/part04/sect_I.4.html`.

```matlab
% Values in sopInstanceMap will be java objects of class
% icr.etherj.dicom.impl.DefaultSopInstance
sopInstanceMap = containers.Map('KeyType','char','ValueType','any');

% Values in seriesMap will be structures with fields for the series
% description and a map of the SOP instances in that series.
seriesMap = containers.Map('KeyType','char','ValueType','any');

% Nested loops to delve into DICOM hierarchy.
for iP = 1:jPatientList.size

    % Note the -1 because Java is zero-indexed.
    jPatient = jPatientList.get(iP-1);

    jStuList = jPatient.getStudyList();
    for iSt = 1:jStuList.size
        jStudy = jStuList.get(iSt-1);
        jSerList = jStudy.getSeriesList();
        for iSe = 1:jSerList.size
            jSeries = jSerList.get(iSe-1);
            jInstList = jSeries.getSopInstanceList();
            nInst = jInstList.size;

            % Initialise structure for this series.
            % map field will be populated in next loop.
            % Note use of char() to convert Java string to MATLAB
            % character array.
            seriesValue = ...
                struct('seriesDescription',char(jSeries.getDescription),...
                'map',containers.Map('KeyType','char','ValueType','any'));

            for iIn = 1:nInst
                jInst = jInstList.get(iIn-1);

                % Only include MR image data.
                if strcmp(char(jInst.getSopClassUid),mrSopClassUid)
                    seriesValue.map(char(jInst.getUid)) = jInst;
                    sopInstanceMap(char(jInst.getUid)) = jInst;
                end

            end

            % Put structure made to represent this series into seriesMap
            seriesMap(char(jSeries.getUid)) = seriesValue;

        end
    end
end

end
```

The following snippet demonstrates use of the utility functions introduced above (**scanPath** and **makeUIDmaps**) to access a single SOP instance (the 12th instance) at the MATLAB command line. The (truncated) output result is shown for calling the **display** method of the variable **jInst**.

```
>> jPatientList = scanPath('/Users/jsmith/DICOM');
>> [sopInstanceMap,seriesMap] = makeUIDmaps(jPatientList);
>> sopInstUIDs = sopInstanceMap.keys;
>> jInst = sopInstanceMap(sopInstUIDs{12});
>> jInst.display;

icr.etherj.dicom.impl.DefaultMrSopInstance
  * File: /Users/jsmith/DICOM/IM_0012.dcm
  * Modality: MR
  * InstanceNumber: 12
  * NumberOfFrames: 1
  * Uid: 1.3.6.1.4.1.9590.100.1.2.129855443013795217323100303
  * SopClassUid: 1.2.840.10008.5.1.4.1.1.4
  * SeriesUid: 1.3.6.1.4.1.9590.100.1.2.406109895141725729822
  * StudyUid: 1.3.6.1.4.1.9590.100.1.2.423723974029532266314
...
```

It should be noted that even for very large data sets, the maps generated by `makeUIDmaps` have a low memory overhead since they contain references to the Java objects corresponding to the DICOM hierarchy and, as mentioned earlier, these objects do not contain the DICOM pixel data, which are loaded on-demand. Although additional DICOM support functions have been introduced in recent years, such as `dicomCollection` in MATLAB R2017b, these do not cover the whole DICOM formalism and are slower than the code discussed above. The execution time using `scanPath` and `makeUIDmaps` is more than an order of magnitude faster than using native MATLAB functionality. These two functions executed on a folder containing 1,873 DICOM files (MR images, total size 358 MB) took approximately 2 seconds to execute on a standard laptop computer, compared with 79 seconds using MATLAB'S `dicomCollection`.

The function `makeUIDmaps` also returns a map with elements corresponding to series instances. This can be used to retrieve image volume data (as shown later), but can also be used to display the series instance UIDs detected and their corresponding series descriptions. We do this below by making use of MATLAB's `cellfun` function[8]. Again, the output is truncated.

```
>> dispFun = @(x,y) disp([x ' | ' y.seriesDescription]);
>> cellfun(dispFun,seriesMap.keys,seriesMap.values,'UniformOutput',false)

1.3.6.1.4.1.9590.100.1.2.111868885943279396573190248 | localizer
1.3.6.1.4.1.9590.100.1.2.73982744034873067922509662 | t1_fl2d_tra_mbh_pat2
1.3.6.1.4.1.9590.100.1.2.134977084202409661820448785 | t2_tse_tra_mbh_pat2
1.3.6.1.4.1.9590.100.1.2.70117251331350751926680744 | t1_fl2d_cor_mbh_pat2
...
```

Reading image data into MATLAB variables

The listing below shows how to get the pixel data and their real-world coordinates from a Java `SopInstance` object by accessing various tag values in the corresponding DICOM file. Three alternative methods of accessing the values are presented:

1. Using EtherJ *get* convenience functions, which are available for a subset of the most commonly used DICOM tags;

2. Using lower-level EtherJ *get* functions for specific data types (float, short, etc.) and using the dcm4che `Tag` function to specify which tag to retrieve. This function uses

---

[8]`cellfun` takes a user-specified function as an argument and applies it to each cell in a cell array.

human-readable syntax, and includes a larger selection of DICOM tags than the EtherJ convenience methods.

3. DICOM tag access via the HEX tag code —this syntax can be used for retrieving any DICOM tag, including private fields.

Note that after executing `scanPath` and `makeUIDmaps` to return `sopInstanceMap` and defining an SOP instance `JInst`, there is a useful shortcut to find a list of the *get* functions, for example, `getRowCount()`. On MATLAB's command line, type "`jInst.`" and then press the *tab* key.

```
% Get pixel data and pixel positions for a SOP instance.

function [pixelData,pixelPos] = getSliceData(jInst)

% Import package to enable human readable DICOM Tag access.
import org.dcm4che2.data.Tag

% Access these metadata tags using EtherJ convenience functions.
nRows = jInst.getRowCount;
nCols = jInst.getColumnCount;
imPos = jInst.getImagePositionPatient;
imOri = jInst.getImageOrientationPatient;
pixSz = jInst.getPixelSpacing;

% Get DICOM Object for accessing other metadata tags.
jDcm = jInst.getDicomObject();

% Access these metadata tags using basic EtherJ functions and
% dcm4che Tag convenience function.
rsSlope = jDcm.getFloat(Tag.RescaleSlope);
rsInt = jDcm.getFloat(Tag.RescaleIntercept);

% Access pixel data using HEX field code.
pixelData = jDcm.getShorts(uint32(hex2dec('7fe00010')));

% Reshape and rescale (if appropriate) the image data.
pixelData = reshape(pixelData,nCols,nRows)';
if rsSlope == 0
    pixelData = double(pixelData);
else
    pixelData = rsSlope*double(pixelData) + rsInt;
end

% Calculate co-ordinates of every pixel - oblique slices are handled correctly.
T = [imOri(1:3)*pixSz(2) imOri(4:6)*pixSz(1) imPos];
[pixIdxRows,pixIdxCols] = ndgrid(0:nRows-1,0:nCols-1);
pixIdx = [pixIdxCols(:)'; pixIdxRows(:)'; ones(1,nRows*nCols)];
pixCoord = T*pixIdx;

% Reshape co-ordinates and append to output variable.
pixelPos.X = reshape(pixCoord(1,:),nRows,nCols);
pixelPos.Y = reshape(pixCoord(2,:),nRows,nCols);
pixelPos.Z = reshape(pixCoord(3,:),nRows,nCols);

end
```

A DICOM series will typically consist of a collection of SOP instances that constitute an image volume when stacked together appropriately[9]. The code listing below shows how to use the functions already presented to read in such a volume from a DICOM series and package it into a 3D array. Such an array could then be analyzed, or visualized using an in-built MATLAB function (e.g., imshow or sliceViewer).

```matlab
% Get series UIDs from map
seriesUIDs = seriesMap.keys;

% Get SOP instances for e.g. 3rd series
jInstMap = seriesMap(seriesUIDs{3}).map;
thisSeriesSopInstUIDs = jInstMap.keys;

% Get sizes from first SOP instance to initialise variables
jInstInit = jInstMap(thisSeriesSopInstUIDs{1});
nRows = jInstInit.getRowCount;
nCols = jInstInit.getColumnCount;
pixelData = zeros(nRows,nCols,jInstMap.Count);
instanceNumber = zeros(1,jInstMap.Count);

% Read pixel data and get instance number for all SOP Instances
for n = 1:jInstMap.Count
    jInst = jInstMap(thisSeriesSopInstUIDs{n});
    instanceNumber(n) = jInst.getInstanceNumber;
    pixelData(:,:,n) = getSliceData(jInst);
end

% Assume slices should be stacked in instance number order
[~,idx] = sort(instanceNumber);
pixelData = pixelData(:,:,idx);
```

18.3   RUNNING MATLAB IN A CONTAINER

18.3.1   The need for containerization

MATLAB is sometimes viewed as a "prototyping environment" in which "scientists play with data", the implication being that once a useful algorithm is created, "real software developers" will program it "properly" with optimized code in a language such as C++. This argument is something of a caricature, failing, as it does, to acknowledge that some scientists do create well-architected solutions and that numerous optimizations are already built into MATLAB. However, it is nonetheless true that when a new technique is deployed in the field—for example, the CT image reconstruction in Ref. [156]—the need often arises to run software within a different environment from the one in which it was developed. For complex applications, this is potentially highly problematic: a rigorous testing regime has to be implemented to ensure that previous results are replicated; re-coding in a different language and computer architecture might yield slightly different values because of numeric precision issues; software dependencies (i.e., library functions that are called) may not exist in the new language, meaning that more has to be rewritten than simply the original application; and licensing might also be an issue.

Containerization is a way of encapsulating all the parts of an application—executables, libraries and other dependencies, settings, system libraries and system tools—in a portable

---

[9]This is not always the case, especially for MR data where multiple volumes may be present in a single series. In some cases (e.g., navigators) the volumes may have different co-ordinates and orientations.

form that can be executed on any target system with a suitable container engine. Named after the ubiquitous shipping containers, which revolutionized trade by sea, containers provide a standardized way of packaging the entire environment of an application so that it will run everywhere. The application inside the container is isolated from the rest of the host system, and all dependencies, such as the *MATLAB Runtime* are built-in, removing the need to install these on the host. A further advantage of this paradigm is that it is easy to construct a well documented catalogue of containers and this can provide an immutable and archivable record of processing methodology.

Containerization is generally regarded as a "lightweight" solution, compared to the related technique of virtualization which creates an entire virtual machine, emulating both hardware and OS, to run a given piece of software. A container stores only the application's dependencies and all containers share the host's own OS kernel to run the encapsulated application instead of needing an OS per application. However, when running cross-OS containers, for example, Linux-based containers on a Windows host, an embedded virtual machine is required to provide the expected kernel interface for the containers thus adding some overhead compared to native containerization. Unfortunately, the optimization of storing only dependencies cannot currently be extended to the trimming of the significant portion of MATLAB functionality that is not needed for any given application, and, as a result, our MATLAB container weighs in at several GB.

18.3.2 Issues encountered in creating a MATLAB container

Linux Zombie Processes

The environment inside a container is not a full OS and this can present problems for software that operates based on that assumption. Under Linux, a process which terminates is eliminated or "reaped" by its parent process and any terminated, but not yet eliminated, process is known as a "zombie". Where the parent of a process is missing, for example, by a user killing it, exiting after an error or by forking a daemon, the special *init* process adopts the orphaned process and becomes responsible for reaping. Linux assumes there is always such an *init* process, but this may not be true in a container. When the task is finished it can become an "unreaped zombie" that persists and consumes kernel resources. In extreme cases this can lead to resource exhaustion. For this reason we used Phusion's *baseimage* container (which creates just such an *init* process), as the foundation for all our containers[10].

OS and Compiler Compatibility

Our work starts with the base Linux container created for *Docker*, one of the most well known container platforms. To this we add the *MATLAB Runtime* and an executable built with the *MATLAB Compiler* on a compatible OS. A MATLAB executable compiled on Microsoft Windows will not run inside a Linux-based container: therefore a *bona fide*, licensed MATLAB installation on a Linux host will be needed here, if the reader's work is normally conducted in Windows. We assume the reader has installed Docker on their host Linux machine[11]. There is no need to download the base container *manually* as it is available in the Docker Registry. The file below—called "Dockerfile"—creates our base MATLAB container "mcr2018b-phusion"[12].

---

[10]See: `https://github.com/phusion/baseimage-docker`.

[11]For installation instructions, see: `https://docs.docker.com/engine/install`

[12]It is possible to change the MATLAB version used in the dockerfile. We also successfully tested with R2019b. However, in addition to changing the environment variable `MATLAB_VERSION` in the docker file, the version number must be changed in `MATLAB_LD_LIBRARY_PATH` (from v95) and the URL for the *MATLAB Runtime* download must be verified as appropriate.

```
# Use Phusion's baseimage to get a proper init process, syslog and others
FROM phusion/baseimage:latest

# Use baseimage-docker's init system.
CMD ["/sbin/my_init"]

# Update everything
RUN apt-get update && apt-get upgrade -y -o Dpkg::Options::="--force-confold"

# Install needed packages
RUN apt-get install wget unzip
RUN apt-get update && apt-get install -y libxt6 xorg xvfb

# Set latest MATLAB version
ENV MATLAB_VERSION R2018b

# Pull and install MATLAB Compiler Runtime
RUN mkdir /mcr-install
RUN wget -P /mcr-install http://www.mathworks.com/supportfiles/downloads/ \
    ${MATLAB_VERSION}/deployment_files/${MATLAB_VERSION}/installers/ \
    glnxa64/MCR_${MATLAB_VERSION}_glnxa64_installer.zip
RUN unzip -d /mcr-install /mcr-install/MCR_${MATLAB_VERSION}_glnxa64_installer.zip
RUN /mcr-install/install -mode silent -agreeToLicense yes

# Cleanup after installer
RUN rm -Rf /mcr-install

# Store the LD_LIBRARY_PATH required by MATLAB but don't set LD_LIBRARY_PATH
# as this breaks apt-get
ENV MATLAB_LD_LIBRARY_PATH /usr/local/MATLAB/MATLAB_Runtime/v95/runtime/ \
    glnxa64:/usr/local/MATLAB/MATLAB_Runtime/v95/bin/glnxa64:/usr/local/ \
    MATLAB/MATLAB_Runtime/v95/sys/os/glnxa64

# Clean up APT when done.
RUN apt-get clean && rm -rf /var/lib/apt/lists/* /tmp/* /var/tmp/*
```

The base container can be created by running the dockerfile from the directory in which it is saved. Open up a terminal, navigate to the directory and enter the following command[13]:

```
docker build -t mcr2018b-phusion:book .
```

The name "book" placed after the colon is an optional tag.

JAR Dependencies for MATLAB Runtime

Correct functioning of compiled MATLAB code that uses Java libraries needs these dependencies to be present in the environment of the executable. While there are various ways to organize the compilation, we found it easiest to add both any additional .m files and any .jar files required using the `-a` option of the `mcc` command[14]. All these dependencies are then

---

[13]The "." at the end of the command indicates that Docker should look in the current directory. For this command syntax to work the dockerfile must be named "Dockerfile". You may require root privileges to run docker on your system. In that case each docker command should be prefaced by, for example, *sudo*, that is, `sudo docker build -t mcr2018b-phusion:book .`

[14]The `mcc` command is a command line alternative to the Application Compiler workflow introduced in Chapter 5.

bundled into the generated executable. In the case of our MATLAB radiomics container this was achieved using a command of the following form:

```
mcc -m radiomicsContainer.m -a etherPackRoot -a radiomicsPackRoot -a javaDependRoot
```

where `radiomicsContainer` is the MATLAB function to be containerized (see Section 18.4.1), `etherPackRoot` and `radiomicsPackRoot` specify the locations of MATLAB packages[15] used by the code and `javaDependRoot` specifies the location of the Java dependencies. The `mcc` command is run at the command line of a terminal on our Linux host and generates a number of files, of which the two most important are `radiomicsContainer` (the executable itself) and `run_radiomicsContainer.sh` (which is used to run the executable).

18.4 EXAMPLE PROBLEM FOR CONTAINERIZATION

18.4.1 Example application: radiomics

It is well recognized that we are not currently making the most of the wealth of radiology data available. Simple image-based measurements, such as the longest diameter across a tumour [157], fail to capture subtleties of shape and image texture. *Radiomics* is the name given to a data-processing paradigm that extracts quantitative imaging "features" from pixel data and analyses them using statistical and machine learning methods similar to those previously developed in other "-omics" disciplines. Promising initial results (e.g., [158]) suggest that such analyses may allow us to discover patterns and associations that are not obvious to radiologists and find features that distinguish disease subtypes. In the clinic, we might be able to use radiomics to stratify patients between different treatment pathways, and process whole tumour volumes, rather than rely on potentially unrepresentative biopsies.

The example presented in the remainder of this chapter should be seen as an overview to our solution to a real-world radiomics problem. For further nuances and more detailed step-by-step instructions, the reader is referred to this book's software repository.

18.4.2 Design and build of the MATLAB radiomics container

The first task in radiomics analysis is to generate the imaging features that will be analysed. Although the work could have been performed using other languages, such as Python, we chose MATLAB for this task for a combination of reasons:

1. MATLAB is an excellent general-purpose image-processing tool.

2. It contains well curated functions to perform specific tasks. We did not want to re-invent these, or rely on online libraries whose provenance and testing might have been uncertain. Functions that were of particular utility for the radiomics use case included: `delaunayTriangulation`, `discretize`, `histcounts` and `histcounts2`, `histogram`, `isosurface`, `poly2mask` and `wfilters`.

3. MATLAB's array-based operations simplify the expression of algorithms greatly compared with some traditional programming languages.

4. Developer familiarity and the ease of use of the Integrated Development Environment, speed up the coding significantly.

---

[15]In MATLAB, *packages* are folders that can contain units such as class folders, functions, and class files. Packages provide namespaces to conveniently organize classes and functions.

A "bare-bones" implementation of the radiomics source code is given in the extracts below. It consists the function radiomicsContainer that calls another function called radiomicsFeatures to loop through the DICOM studies, series and instances. The functions depend on two custom MATLAB packages (ether and radiomics) and two MATLAB classes to handle input and output of data (RadiomicsFeaturesDataSource and RadiomicsFeaturesOutputResults).

The implementation combines many of the ideas referred to earlier in the chapter to parse the data supplied by the user of the container, which are assumed to sit in a top-level directory "/input". The MATLAB code loads all ROIs under the "/input" directory and runs the radiomics feature generation code on them, writing results to the "/output" directory. If the user wishes to test the code outside of a container, for development purposes, the variable USE_MOCK should be set to a value of 1 and the variables MOCK_INPUT and MOCK_OUTPUT set to valid paths on the user's system.

The *scientific* content of the plugin is contained within a method called textureAnalyser.analyse(). The textureAnalyser object is an instance of the TextureAnalyser3D class of our radiomics package. Further discussion of the scientific content of textureAnalyser.analyse() and of the radiomics package is beyond the scope of this chapter. Note that RadiomicsFeaturesDataSource() and RadiomicsFeaturesOutputResults() are also used in the code below and are constructor methods for the MATLAB classes. These classes encapsulate the complexity of supplying data and saving large numbers of results respectively. This simplifies the code listing considerably.

```
function radiomicsContainer()

USE_MOCK = 0;
logger = ether.log4m.Logger.getLogger('radiomicsContainer');

logger.info('Starting radiomics analysis ...');

if USE_MOCK
    MOCK_INPUT  ='<DIR>/radiomics_MATLAB_dev/radiomics_container_mock_data/input';
    MOCK_OUTPUT ='<DIR>/radiomics_MATLAB_dev/radiomics_container_mock_data/output';
    radiomicsFeatures(MOCK_INPUT, MOCK_OUTPUT);
else
    radiomicsFeatures('/input/', '/output/');
end

logger.info('Radiomics analysis complete - container exiting.');

end

function radiomicsFeatures(input, output)

    jAimTk = icr.etherj.aim.AimToolkit.getToolkit();
    jDcmTk = icr.etherj.dicom.DicomToolkit.getToolkit();

    jAimPs = jAimTk.createPathScan();
    jDcmPs = jDcmTk.createPathScan();

    jAimR  = icr.etherj.aim.AimReceiver();
    jDcmR  = icr.etherj.dicom.DicomReceiver();
```

(code continues on next page)

```
jAimPs.addContext(jAimR);

jDcmPs.addContext(jDcmR);
jDcmPs.scan(input, true);
jPr = jDcmR.getPatientRoot();

jiacList = jAimR.getImageAnnotationCollections();

radItemList = ether.collect.CellArrayList('radiomics.RadItem');

for i = 1:jiacList.size()
    % Note how everywhere we have a MATLAB Java object, we need to adjust
    % the loop variables to be zero indexed.jiac = jiacList.get(i-1);
    iac = ether.aim.ImageAnnotationCollection(jiac);
    jiaList = jiac.getAnnotationList();
    for j = 1:jiaList.size()
        jia = jiaList.get(j-1);
        ia = ether.aim.ImageAnnotation(jia);
        iaItem = radiomics.IaItem(ia, iac);
        radItemList.add(iaItem);
    end
end

jPatList = jPr.getPatientList();
for p = 1:jPatList.size()
    jStudyList = jPr.getPatientList().get(p-1).getStudyList();
    for j = 1:jStudyList.size()
        jStudy = jStudyList.get(j-1);
        if strcmp(jStudy.getModality, 'RTSTRUCT')
            jSeriesList = jStudy.getSeriesList();

            for k = 1:jSeriesList.size()
                jSeries  = jSeriesList.get(k-1);
                jSopList = jSeries.getSopInstanceList();

                for m = 1:jSopList.size()
                    jSopInst = jSopList.get(m-1);
                    jRts = jDcmTk.createRtStruct(jSopInst.getDicomObject());
                    rts  = ether.dicom.RtStruct(jRts);
                    jSsrList = ...
                        jRts.getStructureSetModule().getStructureSetRoiList();

                    for n = 1:jSsrList.size()
                        rtRoi = ether.dicom.RtRoi(jSsrList.get(n-1), rts);
                        radItemList.add(radiomics.RtRoiItem(rtRoi, rts));
                    end
                end
            end
        end
    end
end
```

(code continues on next page)

```
    dataSource = RadiomicsFeaturesDataSource(jPr);

    textureAnalyser = radiomics.TextureAnalyser3D();
    % The TextureAnalyser3D object is used in other contexts where explicit
    % setting of the project is needed. Here the data have all been
    % supplied in the /input directory already and this version of the
    % dataSource does not require a projectId to be passed.
    resultList = textureAnalyser.analyse(radItemList.toCellArray(),...
        dataSource, 'DUMMY_PROJECT');
    featureOutput = RadiomicsFeaturesOutputResults(resultList, output);
    featureOutput.sendToFile();

end
```

18.4.3   Deployment: running the container on another computer

As we observed above, our container expects to see its input data in the "/input" directory. How do we get it there?

"Shrink-wrapping" demo data into an image of a Docker container

The files radiomicsContainer and run_radiomicsContainer.sh that are generated by running the mcc command (see Section 18.3.2) are placed in a sub-directory called "compiled_MATLAB" adjacent to another dockerfile, again called "Dockerfile". They are then incorporated into the built Docker image radiomicscontainer-mcr2018b-phusion by running the command:

```
docker build -t radiomics_container-mcr2018b-phusion:book .
```

The dockerfile used to create radiomics image from the base MATLAB image is given below. The first COPY statement copies a required configuration file ("log4m.xml") from the user's system to the docker image. The second COPY statement copies the contents of "compiled_MATLAB" to the required place in the image ("/mnt"). The last COPY statement "hard-wires" the demo data set into "/input" in the image.

```
FROM mcr2018b-phusion:book

# Ether logger configuration
COPY ./log4m.xml /.ether/log4m.xml

# Copy radiomicsContainer standalone MATLAB application.
COPY ./compiled_MATLAB /mnt

# Mock data. Comment this out in production, as data will be loaded into the
    container
# in an appropriate fashion by the calling application. Other modifications may be
# needed in order to set up appropriate shared directories.
COPY ./radiomics_container_mock_data /

# Use mcr to run application.
CMD ["/sbin/my_init", "--", "/mnt/run_radiomicsContainer.sh", \
    "/usr/local/MATLAB/MATLAB_Runtime/v95"]
```

Depending on where the reader wishes to store or publish the image, the reader then uses the **docker export** and **docker import** commands or **docker push** and **docker pull** to retrieve the built image on a different computer running Docker (and which need not have MATLAB installed). On this new host, the container is run on the demo data using:

```
docker run -it radiomics_container-mcr2018b-phusion:book
```

The `-it` flag to `docker run` attaches an interactive terminal to the container. Console output from `fprintf` and other such MATLAB commands are then captured and displayed to the user on their host system running the container.

Mounting new data

Of course, the utility of creating a MATLAB container is to run the code on an arbitrary set of data. A simple way of achieving this is to employ what Docker terms a *volume*. You can create one of these with a command like this:

```
docker create volume myVol1
```

and you can find the directory on the Docker host machine to which this corresponds using the command:

```
docker volume inspect myVol1
```

Typically, this will give you a *mountpoint* that looks something like:

```
/var/lib/docker/volumes/myVol1/\_data
```

In the case of the radiomics example, one would copy the DICOM images on which one wants MATLAB to operate into this directory. Finally, the volume is *mounted* into the image—this time built without the demo data—using:

```
docker run --mount source=myVol1,target=/input \
    radiomicscontainer-mcr2018b-phusion:book
```

So far, nothing has been said regarding where *output* from the container goes. Referring back to the code for the `radiomicsFeatures` function, we see the code `featureOutput.sendToFile()`. Our implementation arranges for MATLAB to place the calculated radiomics features in the "/output" directory of the container. Docker allows multiple mounts on the same image and it is a natural extension of the above to define a second volume with "/output" as its target–the feature data is then preserved on the host system after the container completes its execution and stops. In our example above, however, only the output information displayed to screen in the container is presented to the user on their host system.

Providing data dynamically

It will be readily appreciated that the previous section takes us only so far. It is still necessary to provide the contents of the mount point to the container. Why is this a better option than simply running MATLAB in standard configuration and having the user choose their data using MATLAB's file browser tool? The answer lies in the combination of the container with sophisticated third party tools, whose function is to "serve up" data dynamically to the mount point. In the final section of the chapter, we explore one such system.

18.5 XNAT: ORCHESTRATING THE IMAGE ANALYSIS OF LARGE PATIENT COHORTS

In order to deploy the radiomics code for use at the scale needed for routine processing of clinical data, one needs to integrate the MATLAB container properly into the research radiology workflow. XNAT (the eXtensible Neuroimaging Archive Toolkit)[159] is now gaining significant traction as an image data management system in many areas of biomedicine, well beyond its origins in neuroscience. Designed to address the issues of archiving and accessing large quantities of DICOM image data in a research context, it is, in many ways, the academic equivalent of a hospital PACS.

XNAT is made up of three main components: a structured file system (archive); a database to allow files in the archive to be easily searched; and a user interface in the form of a web application. Within the archive, data are conveniently organized in a hierarchy: projects, subjects, experiments (e.g., imaging visits), scans (the data coming from the scanners) and assessors (activity analysing the experimental data).

The XNAT Container Service provides a highly flexible interface between the data stored in the archive and an arbitrary set of containers imported into XNAT. Although the technical detail is beyond the scope of this chapter, the practical upshot is that the Container Service manages what goes into "/input" at the time of a container's launch. "/input" thus "points at" the XNAT data archive at any desired level of the hierarchy, allowing the container to "see" and manipulate an appropriate subset of the stored data. Similarly, the Container Service takes whatever is written to "/output" and handles all the complexities of deciding where to put it and how to update the database. This represents an extremely powerful layer of abstraction and allows algorithm designers to encapsulate their data processing *without any need to know the details about how or where those data are really stored, or anything about the security credentials required to access those data or write results back.*

18.6 CONCLUSION

MATLAB provides highly tested functionality with a development platform that is popular with scientists. The integration of libraries written in other languages (particularly Java) and the containerization paradigm allow us to experiment with different processing algorithms, focusing scientific activity on the truly novel parts of a project, while leaving an external informatics infrastructure to take care of the application of the algorithms to large-scale real-world data.

> MATLAB toolboxes used in this chapter:
> *MATLAB Compiler*
> *Wavelet Toolbox*
> *Image Processing Toolbox*
>
> Index of the in-built MATLAB functions used:

cellfun	double	poly2mask	wfilters
char	histcounts	reshape	zeros
containers.Map	histcounts2	sort	
delaunayTriangulation	histogram	strcmp	
discretize	isosurface	struct	

## CHAPTER 19

Estimation of arterial wall movements

Magnus Cinthio, John Albinsson, Tobias Erlöv

Department of Biomedical Engineering, Faculty of Engineering, Lund University, Lund, Sweden

Tomas Jansson

Biomedical Engineering, Department of Clinical Sciences, Lund University, Lund, Sweden
Clinical Engineering Skåne, Region Skåne, Sweden

Åsa Rydén Ahlgren

Clinical Physiology, Department of Translational Medicine, Lund University, Skåne University Hospital, Malmö, Sweden

CONTENTS

I N THIS CHAPTER we will describe an intriguing biomedical research project where MAT-LAB has been invaluable for its implementation–both in its initial phase, when developing the algorithms and the graphical user interface and when other research groups started to use the methods. For the project, we needed to develop methods that could track two-dimensional (2D) movements in ultrasound cine loops. In this chapter we will describe part of our development process with lessons of general applicability to motion measurement.

19.1 THE LONGITUDINAL MOVEMENT OF THE ARTERIAL WALL

In cardiovascular research the diameter change of the arterial wall has for a long time been the subject of extensive interest. Measurement of the diameter change is now an established tool in cardiovascular research forming the basis for estimation of arterial wall stiffness, an important factor for the performance of the heart. In contrast, the longitudinal movement, that is, along the arterial wall, has gained little attention and it was earlier assumed that the longitudinal movement was negligible compared to the diameter change[160]. However, using an in-house developed ultrasonic measurement technique on large arteries *in vivo*[161], we demonstrated the presence of a distinct longitudinal movement of the innermost layers of the arterial wall, having the same magnitude as the diameter change[162]. Figure 19.1 shows an example of the appearance of estimated 2D movement loops of the inner walls of the common carotid artery in a 45-year-old female. The function of the longitudinal movement are at present largely unknown. However, among other things we have shown that our stress hormones adrenaline and noradrenaline seem to markedly influence the longitudinal movement, thereby constituting a possible link between mental stress and cardiovascular disease [163]. Further, changes in longitudinal movement has been reported to be associated with atherosclerosis [164] and risk factors for cardiovascular disease [165, 166].

Measurements of the arterial wall movement are challenging. For large arteries, the wall is around 1 mm thick and consists of three main layers. The sub-millimeter movement has a velocity up to 30 mm/s and can change direction several times during a cardiac cycle. This puts high demands on both the spatial and temporal resolutions of the measurement method. We have found two-dimensional high-frequency diagnostic ultrasound (10–15 MHz) B-mode imaging to be the best alternative as it provides sub-millimeter spatial resolution and millisecond time resolution. Even so, measurements of the longitudinal movement are especially challenging as the motion occurs perpendicular to the ultrasound beam (in the lateral direction) where the spatial resolution is limited. When the vessel is scanned longitudinally the near and far arterial walls appear as a double line pattern on both sides of the dark lumen (see Figure 19.1). It is important that the walls are visible during several cardiac cycles to be able to track the movement. When this condition is fulfilled, we have found 2D block matching to be useful to track the arterial wall movement both in axial and lateral directions in the ultrasound cineloop.

To provide visual feedback during the arterial wall movement estimation, we prefer to use Lagrangian motion description instead of the more commonly used Eulerian motion description. The rationale behind this is that all motion estimations include an error term. Even if each estimation error is small, many small errors can accumulate to a larger one. Lagrangian motion description of a region of interest estimates its position throughout a cine loop and the visual inspection of the estimated position plotted in the cine loop reveals if the accumulated error becomes too large.

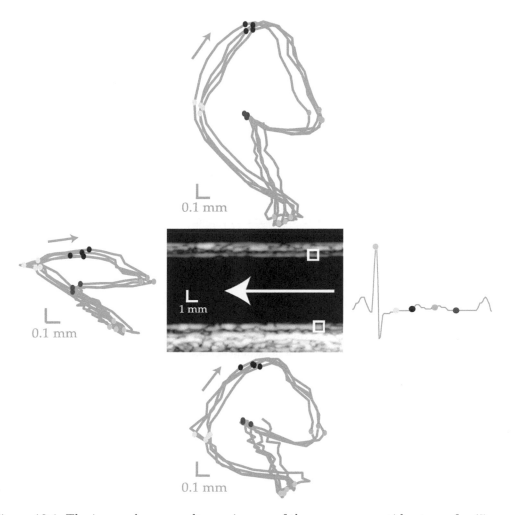

Figure 19.1: The image shows an ultrasonic scan of the common carotid artery of a 45-year-old female. The white squares mark the measurement positions. The direction of the blood flow and an antegrade longitudinal movement is indicated by the white arrow. The magnified 2D movement loops (*orange lines*) of the regions of interest in the innermost layers of the arterial wall are shown above and below the ultrasound image. Both the recorded radial and longitudinal movements are of millimeter magnitude. The radial movement take place in axial direction in the ultrasound image whereas the longitudinal movement take place in the lateral direction. The longitudinal movement of the far wall versus the differential movement between the near and far walls in radial direction (the diameter change) of the artery is shown to the left. The orange arrows indicate the direction of the movement. The ECG-trace and the colored dots to the right indicate the timing of the movement, running throughout four consecutive cardiac cycles.

19.2   BLOCK MATCHING

One general method to measure movements in digital cineloops is to estimate the similarity between two image blocks in two consecutive frames using block matching. A block is a matrix consisting of $m \times n$ pixels in a 2D image or $m \times n \times q$ voxels in a 3D image. In block matching, a reference block, often also called kernel, in frame k is compared with several blocks in the search region in the consecutive search frame $k+1$ to determine which block in frame $k+1$ is most similar to the kernel obtained in frame k. To determine the similarities between two image blocks for example cross correlation, normalized cross correlation or sum of absolute difference (SAD) can be used. The 2D versions of cross correlation and normalized cross correlation are in-built in MATLAB in the functions **xcorr2** (the *Signal Processing Toolbox*) and **normxcorr2** (the *Image Processing Toolbox*), respectively. The 2D version of SAD, that is an exhaustive search within a given search region to find the match position that minimizes the SAD, can be implemented by:

```
function [bestMatchPos, SADvalue] = SAD(kernel, imageData, n, m, xPos, yPos)
% (kernel) and (imageData) are two 2D image matrices. (n) and (m) state the ...
% size of the search region in (imageData), and (xPos) and (yPos) the centrum ...
% position of the search region. (SADvalue) returns similarity coefficients ...
% that are important in sub-pixel estimations, and (bestMatchPos) returns the ...
% coordinates of the position with the best match.

[ySizeKernel, xSizeKernel] = size(kernel);
ySize = floor(ySizeKernel./2); xSize = floor(xSizeKernel./2);
SADoutput = zeros(n*2+1,m*2+1);
for i = -n:n
    for j = -m:m
        block = imageData(yPos+i-ySize:yPos+i+ySize,xPos+j-xSize:xPos+j+xSize);
        SADoutput(i+n+1,j+m+1) = sum(sum(abs(kernel-block)));
    end
end

% Find the position of the best match
SADtemp = SADoutput(2:end-1,2:end-1);
[~, minIndex] = min(SADtemp(:));
[r,c] = ind2sub(size(SADtemp),minIndex);
bestMatchPos = [yPos, xPos] + [r,c] + [1,1] - [n+1, m+1];

% Save similarity coeffcients that is needed for sub-pixel estimations
SADvalue = zeros(1,5); SADvalue(1) = SADoutput(r+1,c+1);
SADvalue(2) = SADoutput(r+1,c-1+1); SADvalue(3) = SADoutput(r+1,c+1+1);
SADvalue(4) = SADoutput(r-1+1,c+1); SADvalue(5) = SADoutput(r+1+1,c+1);

end % Function
```

It is possible to save computational cost by using a sparse iterative search algorithm [167]. Adaptive rood pattern search (ARPS) uses a sparse iterative search with a search radius that decreases at each iteration, converging to a solution[167], and can be implemented by:

```
function [bestMatchPos, SADvalue] = ARPS(kernel, imageData, startPos, speed)
% (kernel) and (imageData) are two 2D image matrices. (startPos) state the ...
% centre position of the search region and (speed) state the movement ...
% in the previous estimation. (bestMatchPos) returns the coordinates ...
% of the position with the best match and (SADvalue) returns similarity ...
% coefficients that are important in sub-pixel estimations.

[ySizeKernel, xSizeKernel] = size(kernel);
ySize = floor(ySizeKernel./2); xSize = floor(xSizeKernel./2);
searching = true;
```

(code continues on the next page)

```
% Set the intial search positions
searchRadius = max([1, round(speed(1)), round(speed(2))]);
searchPos = [startPos; startPos(1)-searchRadius startPos(2); ...
    startPos(1)+searchRadius startPos(2); ...
    startPos(1) startPos(2)-searchRadius; ...
    startPos(1) startPos(2)+searchRadius; ...
    startPos + round(speed)];
SADvalues = zeros(1,6);
while searching
    % Estimate the similaritiy coefficients in the search positions
    for i = 1:length(searchPos)
        block = imageData(searchPos(i,1)-ySize:searchPos(i,1)+ySize, ...
            searchPos(i,2)-xSize:searchPos(i,2)+xSize);
        SADvalues(i) = sum(sum(abs(kernel - block)));
    end

    % Find the search position with minimum similaritiy coefficient
    [~, minIndex] = min(SADvalues);
    newPos = searchPos(minIndex,:);

    % Stop the estimation if we used the same search position twice
    if (minIndex == 1) && (searchRadius == 1)
        searching = false;
    else
        searchRadius = max(1, floor(searchRadius/2));
        searchPos = [newPos; newPos(1)-searchRadius newPos(2); ...
            newPos(1)+searchRadius newPos(2); ...
            newPos(1) newPos(2)-searchRadius; ...
            newPos(1) newPos(2)+searchRadius];
        SADvalues = zeros(1,5);
    end
end
bestMatchPos = newPos;
SADvalue = [SADvalues(1); SADvalues(4); SADvalues(5); SADvalues(2); SADvalues(3)];

end % Function
```

The resolution of the motion estimate obtained by block matching depends on the image resolution (pixels per mm). Often this resolution is not high enough to obtain accurate and smooth motion estimates. The most basic way to improve the resolution is to interpolate both the kernel and the search region by a 2D interpolation function prior to matching. In MATLAB such a function is in-built in the function `interp2`. 2D image interpolation provides high quality sub-sample estimations but is very time-consuming. Therefore, alternative methods, such as parabolic [168] and grid slope [169] interpolation or a combination between them have been developed [170]. The methods estimate the sub-sample position from a number of adjacent similarity coefficient values following matching at the normal resolution. The methods separately estimate the lateral and the axial directions in the image.

In MATLAB the parabolic interpolation for the SAD case can be implemented by:

```
function [subX, subY] = PI(SADvalue)
% (SADvalue) contains similarity coefficients of interest and (subX) and ...
% (subY) are the interpolated positions in units of pixels relative to ...
% the minimum pre-estimated position in lateral and axial directions.

% Lateral direction
if (SADvalue(2)+ SADvalue(3) ~= 2*SADvalue(1))
    subX = (SADvalue(2) - SADvalue(3)) / 2 ...
        / (SADvalue(2) + SADvalue(3) - 2*SADvalue(1));
else
    subX = 0;
end

% Axial direction
if (SADvalue(4)+ SADvalue(5) ~= 2*SADvalue(1))
    subY = (SADvalue(4) - SADvalue(5)) / 2 ...
        / (SADvalue(4) + SADvalue(5) - 2*SADvalue(1));
else
    subY = 0;
end

end % Function
```

and a modified grid slope interpolation by:

```
function [subX, subY] = GS(SADvalue, kernel, imageData, bestMatchPos)
% (SADvalue) contains similarity coefficients of interest, (kernel) and ...
% (imageData) are two 2D image matrices, (bestMatchPos) is the position in ...
% (imageData) with the best similarity to (kernel), and (subX) and (subY) are ...
% the interpolated positions in units of pixels relative to the minimum ...
% pre-estimated position in lateral and axial direction.

[ySizeKernel, xSizeKernel] = size(kernel);
ySize = floor(ySizeKernel./2); xSize = floor(xSizeKernel./2);

% Lateral direction
if SADvalue(2) < SADvalue(3)
    normBlock = imageData(bestMatchPos(1)-ySize:bestMatchPos(1)+ySize, ...
        bestMatchPos(2)-1-xSize:bestMatchPos(2)-1+xSize);
else
    normBlock = imageData(bestMatchPos(1)-ySize:bestMatchPos(1)+ySize, ...
        bestMatchPos(2)+1-xSize:bestMatchPos(2)+1+xSize);
end
block = imageData(bestMatchPos(1)-ySize:bestMatchPos(1)+ySize, ...
    bestMatchPos(2)-xSize:bestMatchPos(2)+xSize);
normSADx = sum(sum(abs(normBlock-block)));
if (normSADx == 0) || (SADvalue(2) == SADvalue(3))
    subX = 0;
elseif SADvalue(2) < SADvalue(3)
    subX = -0.5 * (1 - (SADvalue(2) - SADvalue(1))/normSADx);
else
    subX = 0.5 * (1 - (SADvalue(3) - SADvalue(1))/normSADx);
end
```

(code continues on the next page)

```
% Axial direction
if SADvalue(4) < SADvalue(5)
    normBlock = imageData(bestMatchPos(1)-1-ySize:bestMatchPos(1)-1+ySize, ...
        bestMatchPos(2)-xSize:bestMatchPos(2)+xSize);
else
    normBlock = imageData(bestMatchPos(1)+1-ySize:bestMatchPos(1)+1+ySize, ...
        bestMatchPos(2)-xSize:bestMatchPos(2)+xSize);
end
block = imageData(bestMatchPos(1)-ySize:bestMatchPos(1)+ySize, ...
    bestMatchPos(2)-xSize:bestMatchPos(2)+xSize);
normSADy = sum(sum(abs(normBlock-block)));
if (normSADy == 0) || (SADvalue(4) == SADvalue(5))
    subY = 0;
elseif SADvalue(4) < SADvalue(5)
    subY = -0.5 * (1 - (SADvalue(4) - SADvalue(1))/normSADy);
else
    subY = 0.5 * (1 - (SADvalue(5) - SADvalue(1))/normSADy);
end

end % Function
```

The estimated interpolated positions become: yPos = bestMatchPos(1)+subY and xPos = bestMatchPos(2)+subX, where bestMatchPos can be found using SAD or ARPS.

Parabolic interpolation gives a biased estimation, that is, the mean error is not zero, if the sub-pixel displacement is not close to a whole pixel. It is especially notable if the sub-pixel displacements are greater than 0.2 pixels, that is, where abs(subX)>0.2 or abs(subY)>0.2. Grid slope interpolation gives unbiased motion estimates, but the mean error has, on the other hand, a large standard deviation for sub-pixel displacements close to a whole pixel, that is, where abs(subX)$\approx$ 0.0 or abs(subY)$\approx$ 0.0. We therefore proposed a new fast unbiased and stable sub-sample method[170], denoted GS15PI, in which the sub-sample displacement was first estimated using parabolic interpolation. If the absolute sub-sample estimation was larger than a threshold (chosen to be 0.15), the sub-sample estimate was recalculated by grid slope interpolation. The function can be implemented by:

```
function [subX, subY] = GS15PI(SADvalue, kernel, imageData, bestMatchPos)
% (SADvalue) contains similarity coefficients of interest, (kernel) and ...
% (imageData) are two 2D image matrices, (bestMatchPos) is the position in ...
% (imageData) with the best similarity to (kernel, and (subX) and (subY) are ...
% the interpolated positions in units of pixels relative to the minimum ...
% pre-estimated position in lateral and axial directions.

[subX, subY] = PI(SADvalue);
if (abs(subX) > 0.15) || (abs(subY) > 0.15)
    [subX_GS, subY_GS] = GS(SADvalue, kernel, imageData, bestMatchPos);
    if abs(subX) > 0.15
        subX = subX_GS;
    end
    if abs(subY) > 0.15
        subY = subY_GS;
    end
end

end % Function
```

19.3   ARTERIAL WALL MOVEMENT MEASUREMENTS

In our first method to estimate 2D arterial wall movements we tracked an individual ultrasonic echo in the arterial wall through space and time, rather than a group of echoes as in conventional speckle tracking[161]. To accomplish this, we used cross correlation as matching criteria and much smaller search regions (approximately 0.75×0.25 mm) and kernels (approximately 0.07×0.05 mm) than previously had been suggested[161]. This method provided accurate tracking as long as the kernel was locked on the correct echo [171]. However, the method demanded a very high scan quality and a careful choice of which echo to be tracked. To be less sensitive to the placement of the kernel in the arterial wall we adopted a new approach. We used larger kernels, a sparse search pattern algorithm (ARPS minimizing SAD[167]), two kernels from two consecutive frames[167] and the new unbiased sub-pixel estimator GS15PI[170]. The function can be implemented by:

```
function [xPos, yPos] = LMovSpeckletracking(image3D, yKernel, xKernel, ...
    ySizeKernel, xSizeKernel)
% (image3D) contains a 2D ultrasound cineloop, (yKernel) and (xKernel) ...
% the centre position of the kernel and (ySizeKernel) and (xSizeKernel) ...
% the size of the kernel

[~,~,nbrOfFrames] = size(image3D);
ySize = floor(ySizeKernel./2); xSize = floor(xSizeKernel./2);
xPos = zeros(nbrOfFrames,1); yPos = zeros(nbrOfFrames,1);
xPos(1) = xKernel; yPos(1) = yKernel;


% Extract kernel in frame 1 and estimate movement between frame 1 and 2
kernelB = image3D(yKernel-ySize:yKernel+ySize,xKernel-xSize:xKernel+xSize,1);
[bestMatchPos, SADvalue] = ARPS(kernelB, image3D(:,:,2), ...
    [yKernel, xKernel], [0 0]);
[subX, subY] = GS15PI(SADvalue, kernelB, image3D(:,:,2), bestMatchPos);
xPos(2) = bestMatchPos(2) + subX; yPos(2) = bestMatchPos(1) + subY;
for i = 3:nbrOfFrames
    % Save kernel obtained in frame i-2
    kernelA = kernelB;

    % Extract kernel in frame i-1
    kernelPos = [round(yPos(i-1)) round(xPos(i-1))];
    kernelB = image3D(kernelPos(1)-ySize:kernelPos(1)+ySize, ...
        kernelPos(2)-xSize:kernelPos(2)+xSize,i-1);

    % Estimate movement between frame i-1 and frame i
    [bestMatchPos, SADvalue] = ARPS(kernelB, image3D(:,:,i), kernelPos, ...
        [yPos(i-1)-yPos(i-2) xPos(i-1)-xPos(i-2)]);
    [subX1, subY1] = GS15PI(SADvalue, kernelB, image3D(:,:,i), bestMatchPos);
    xPosARPS = bestMatchPos(2) + subX1 - (kernelPos(2) - xPos(i-1));
    yPosARPS = bestMatchPos(1) + subY1 - (kernelPos(1) - yPos(i-1));

    % Estimate movement between frame i-2 and frame i
    [bestMatchPosSAD, SADvalue] = SAD(kernelA, image3D(:,:,i), 3, 3, ...
        bestMatchPos(2), bestMatchPos(1));
    [subX2, subY2] = GS15PI(SADvalue, kernelA, image3D(:,:,i), bestMatchPosSAD);
    xPosSAD = bestMatchPosSAD(2) + subX2 - (round(xPos(i-2)) - xPos(i-2));
    yPosSAD = bestMatchPosSAD(1) + subY2 - (round(yPos(i-2)) - yPos(i-2));
```

(code continues on the next page)

```
    % Estimate average movement
    xPos(i) = (xPosARPS + xPosSAD)/2;
    yPos(i) = (yPosARPS + yPosSAD)/2;
end

end % Function
```

Two example data sets are provided with the digital materials accompanying this chapter: one simulated (*in silico*) and one derived from a human subject (*in vivo*). The MATLAB scripts `insilicoExample`.m and `invivoExample`.m demonstrate the use of the algorithm. The size of a kernel is critical and needs to be evaluated for each application. A larger kernel size often gives a more stable estimate but there are several reasons to minimize the size of the kernel, including improved spatial resolution, decreased computational complexity, improved motion resolution and improved performance under conditions of motion gradients [172].

The arterial wall is relatively thin so it is advisable to use a kernel consisting of only a few pixels in the axial direction while it can be much larger in the lateral direction. The small kernel in axial direction can however make it difficult to track the radial movement of the arterial wall. To overcome this, we first tracked the radial movement separately and used the estimated radial movement as *a priori* information in the block-matching [173, 174]. That step, however, is beyond the scope of this chapter.

19.4   CONCLUDING REMARKS

Block matching is an effective tool to measure 2D motion in all type of digital cine loops. Our implementation is a fast and robust state-of-the-art method that combines a sparse search algorithm, an unbiased sub-pixel estimator and multiple kernels per motion estimate. It is a general method that can be used for any 2D motion estimation purpose. In measurements of the longitudinal movement of the arterial wall, block matching methods can be assisted by first estimating the radial movement, and then using the estimated radial movement as *a priori* information during the block matching.

MATLAB toolboxes used in this chapter:			
None			
Index of the in-built MATLAB functions used:			
abs	floor	max	size
ceil	ind2sub	min	sum
end	length	round	zeros

CHAPTER **20**

Importation and visualization of ultrasound data

Tobias Erlöv, Magnus Cinthio

Department of Biomedical Engineering, Faculty of Engineering, Lund University, Lund, Sweden

Tomas Jansson

Biomedical Engineering, Department of Clinical Sciences, Lund University, Lund, Sweden
Clinical Engineering Skåne, Region Skåne, Sweden

CONTENTS

Raw data can often be exported from medical imaging devices but different manufacturers and models may export in different file formats. Therefore, in order to open such a file in MATLAB one has to create a dedicated script. In this case study we demonstrate an example of how to read an arbitrary, but predefined file format containing ultrasound image data and how to display the ultrasound images.

20.1 INTRODUCTION TO ULTRASOUND DATA

Ultrasound is sound with a frequency above the audible range of the human ear. In medical ultrasound scanners the frequency range is usually in the range of 1–15 MHz. Using an ultrasound transducer, sound pulses are transmitted into the body where they are reflected at various structures, for example, boundaries between different tissues, due to differences in acoustical properties. A fraction of the reflected sound travels back to the transducer and the scanner then measures the time delay from the transmitted pulse to the reception of the different echoes (reflected pulses). By knowing the speed of sound the scanner can calculate the origin of the echoes and thereby, similarly to the bats and dolphins, create a picture of the reflecting structures.

The raw ultrasound data, called Radio Frequency (RF) data, is an oscillating signal describing the sound pressure measured at the transducer face at a time corresponding to the depth where the echo is produced in the image. A linear array ultrasound transducer typically consists of 256 piezo-electric elements placed in a row. A traditional ultrasound image is created line by line, where each line results from a beam-formed combination of pulses from an appropriate aperture (certain number of active piezoelectric elements). Hence, one raw ultrasound image could typically consist of up to 256 columns, where each column describes how the sound pressure varies with time (= depth) in the corresponding tissue segment. The B-Mode (Brightness Mode) image that normally is displayed during medical ultrasound examinations is the amplitude of the sound pressure where brighter pixels correspond to stronger echoes. In other words, the B-Mode image is the amplitude (envelope) of the RF data, usually with some additional post processing. Figure 20.1 shows a partial line of RF data and the corresponding envelope. The B-Mode images can normally be exported as standardized DICOM images. The RF data, which contains more information and is commonly used in research environments, is usually not exportable from clinical scanners and never in DICOM format.

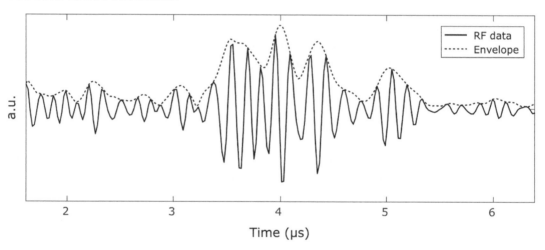

Figure 20.1: Ultrasound raw data (RF data) (*solid*) and its envelope (*dashed*).

20.2   STRUCTURE OF A DATA FILE

Once an ultrasound acquisition is saved in a raw format the resulting data file is often accompanied by one or more files with additional information regarding the specific settings used. The manufacturers can provide information about the file structures for their specific system. Let us define an example binary file structure for this case study. The file "imageExample.rf" is structured according to Figure 20.2.

Figure 20.2: Structure of the file containing ultrasound data.

Table 20.1 shows the first 32 bytes that represents the *file header*.

Table 20.1: Binary representation of the file header.

Size	Description
Unsigned integer (4*1 bytes)	Version number
Unsigned integer (4*1 bytes)	Format of saved data
Unsigned integer (1*4 bytes)	Number of frames in this data file
Unsigned integer (5*4 bytes)	Reserved for future use

Then the file continues frame by frame where the image data in each frame is preceded by a 32 byte *frame header* as described in Table 20.2.

Table 20.2: Binary representation of the frame header.

Size	Description
Double float (1*8 bytes)	Hardware time stamp counted in ms
Unsigned integer (4*1 bytes)	Format of the frame
Unsigned integer (1*4 bytes)	Size in bytes of the frame data (not including header)
Unsigned integer (4*4 bytes)	Reserved for future use

The image data that follows each frame header is stored column-wise and each sample is represented by a float number (4 bytes). The additional files stored with the ultrasound data contain information about which transducer and which settings were used during the acquisition. In this case study we define the file *settingsExample.xml* containing the parameters such as the following:

```
<systemSettings>
        <parameter="Data-Format" value="RF"/>
        <parameter="Study-Name" value="Example Study"/>
        <parameter="Series-Name" value="Series 1"/>
        <parameter="Image-Label" value="Carotid"/>
        <parameter="Study-Owner" value="Operator 1"/>
        <parameter="Acquired-Time" value="10:05:17 AM"/>
        <parameter="Acquired-Date" value="01/10/2018"/>
        <parameter="Transducer-Name" value="Linear"/>
        <parameter="Image-Depth" value="24"/>
        <parameter="Depth-Offset" value="1"/>
        <parameter="Samples" value="5024"/>
        <parameter="Image-Width" value="23"/>
        <parameter="Lines" value="256"/>
        ...
</systemSettings>
```

These can be read into MATLAB through string matching which will be described in the next section.

20.3 READ DATA INTO MATLAB

We start with the file containing the image settings. In this example these were stored in an xml-format for which there is a native MATLAB function **xmlread**. However, in order to generalize this case to cover also other formats, we will treat the file as any other binary file format. To open the file, the function **fopen** is used. This will give us a file identifier that can be used by other functions:

```
fileID = fopen('settingsExample.xml');
```

There are multiple solutions to extract the parameters from the file. One solution could be to read the file item by item, sequentially identifying parameters of interest. This could be beneficial if all parameters are of interest. However, if only a few parameters are of interest and if the file does not contain a huge amount of data one might consider to first read and store all the items at once in a local variable and then close the file. To do so, we read the file with the function **fread** and then close the file with **fclose**. Finally, as our file is a text file, the vector **allBytes** will be stored as a character array:

```
allBytes = fread(fileID);
fclose(fileID);
allSettingsText = char(allBytes');
```

In the variable **allSettingsText** the parameters can be identified using the function **strfind**:

```
parameterOfInterest = 'Study-Owner';

% Return the position of the byte corresponding to the first letter
bytePos = strfind(allSettingsText,parameterOfInterest);
```

In this file structure we notice that the value following each parameter is surrounded by quotation marks (""). More specifically, after identifying the name of the parameter of interest, the corresponding value is situated between the third and fourth quotation mark (e.g., "Study-Owner" value="Operator 1"). Thus the value can be identified by extracting the characters between the second and third quotation mark *after* the parameter name (the very first quotation mark is *before* the parameter name). The function **strfind** is again useful for this purpose:

```
allQuotationsMarks = strfind(allSettingsText,'"');

% Find the first 3 QM following the bytePos
indexFirstThree = find(allQuotationsMarks>bytePos,3,'first');

% Find first byte of value = first byte after second QM
firstByte = allQuotationsMarks(indexFirstThree(2)) + 1;

% Find last byte of value = last byte before third QM
lastByte = allQuotationsMarks(indexFirstThree(3)) - 1;

% Extract value between second and third QM
studyOwner = allSettingsText(firstByte:lastByte);
```

The process is then repeated to extract and calculate all parameters of interest, for example, the number of image rows, columns, depth-offset (called *Samples*, *Lines* and *Depth-Offset* in *settingsExample.xml*) and the number of samples (pixels) per millimeter in both directions (referred to as **pixelsPerMmWidth** and **pixelsPerMmHeight** later in the code)[1]. Next the

---

[1]MATLAB offers other functions that can be useful for tasks like this, such as **extractBetween** for identifying parts of strings or **split** for splitting them based on a specified delimiter.

image data file is opened and the first step is to read the first 32 bytes and save the parameters, if they are of interest. The function **fread** will read the specified number of bytes and then place a file pointer at the current byte location. When calling the same file identifier again, the function will continue from the location of the pointer. Note that different parameters might have been stored with different binary formats (e.g., **unit8** or **double**). This has to be set in the **fread** function:

```
fileID = fopen('imageExample.rf');

%%% Start File Header %%%
dataVer = fread(fileID,4,'uint8'); % Version number
dataFormat = fread(fileID,4,'uint8'); % Format of image data
nFrames = fread(fileID,1,'uint32'); % Number of frames
fread(fileID,5,'uint32'); % Reserved (no storing, read to move pointer only)
%%% End file header %%%
```

Now that the number of frames is known it is appropriate to declare a variable for the ultrasound raw data (RF data) with the correct size. The variable is then filled with data through a **for**-loop:

```
rfData = zeros(nSamples, nLines, nFrames);

for k=1:nFrames
    %%% Start frame header %%%
    timeStamp = fread(fileID,1,'double'); % Time stamp in milliseconds
    if k==1
        startTime=timeStamp;
    end
    if k==nFrames
        stopTime=timeStamp;
    end
    fread(fileID,4,'uint8'); % Frame format (not stored)
    fread(fileID,1,'uint32'); % dataSize (not stored)
    fread(fileID,4,'uint32'); % Reserved (not stored)
    %%% End frame header %%%

    % Reads frame data
    frameData = fread(fileID, nSamples*nLines, 'single');
    % Converts frame data from vector to matrix format
    rfData(:,:,k) = reshape(frameData,nSamples,nLines);
end

fclose(fileID); % Close file

totalTime=stopTime-startTime; % Calculate acquisition time
frameRate = (nFrames-1)/totalTime; % Image frame rate
```

Note that in this example the frame rate was not given in the image settings file but was calculated through the time stamps given in the frame data.

20.4 GENERATING AND VISUALIZING B-MODE IMAGES

This section provides an example of how to generate and visualize a standard B-Mode image from the RF data. The first step in creating a B-Mode image is to extract only the pressure amplitude from the RF data. This is usually achieved through transformation of the RF data into a complex notation, either with quadrature demodulation or using the Hilbert transform. Quadrature demodulation is where the signal is multiplied with two reference signals (separated by 90 degrees of phase) and low-pass filtered to form two orthogonal signals representing the complex notation[175]. With more computational power readily available, the Hilbert transform is increasingly used as it essentially achieves the same result utilizing the Fast Fourier Transform[176]. In this case study we use the Hilbert transform through MATLAB's native function **hilbert** to create the complex notation. The amplitude is then simply the absolute value of the complex notation. There are infinite options regarding post processing of the displayed image, but here we only include a log compression to enhance readability of weaker details.

```matlab
% Display image
hilbertFrame = hilbert(rfData(:,:,1)); % Perform HT on first frame
amplitudeFrame = abs(hilbertFrame); % Extract the amplitude
bmode = log10(amplitudeFrame); % Log compress
clims = [1.5 4]; % Define display range
imageHandle = imagesc(bmode,clims); % Display B-Mode image
colormap(gray) % Change to grayscale

% Define start and stop values in mm for x- and y-directions
startX = 1/pixelsPerMmWidth;
stopX = nLines/pixelsPerMmWidth;
startY = 1/pixelsPerMmHeight + depthOffset;
stopY = nSamples/pixelsPerMmHeight + depthOffset;

% Change the x- and y-ranges for the displayed data
set(imageHandle, 'XData', [startX, stopX]);
set(imageHandle, 'YData', [startY, stopY]);
axis([startX stopX startY stopY])

% Set labels
xlabel('Width (mm)')
ylabel('Depth (mm)')
```

Figure 20.3 below shows the displayed ultrasound image (of a common carotid artery).

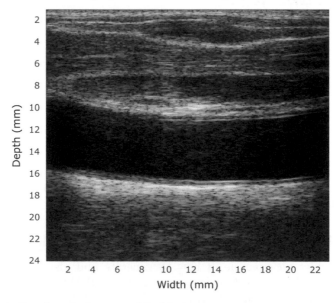

Figure 20.3: Displayed ultrasound B-Mode image of a common carotid artery.

The previous example was used to display a single ultrasound B-Mode image. By adding a for-loop one can choose to play an entire cine loop of acquired ultrasound images[2]. In this case it could be useful to define a figure used for the plotting. This will result in a more stable performance if one handles several figures simultaneously. The gca command is used to get an ID for the current axes which, through imagesc, can be used to force plotting in a specific figure.

```
figure; % Create figure
figureID = gca; % Retrieves ID for current axes
for k=1:nFrames
    hilbertFrame = hilbert(rfData(:,:,k)); % Perform HT on frame k
    amplitudeFrame = abs(hilbertFrame); % Extract the amplitude
    bmode = log10(amplitudeFrame); % Log compress
    imageHandle = imagesc(figureID,bmode); % Display image
    caxis(figureID,[1.5 4]); % Set colormap display limits
    colormap(gray) % Change colormap to grayscale
    % Change the x- and y-ranges for the displayed data
    set(imageHandle, 'XData', [startX, stopX]);
    set(imageHandle, 'YData', [startY, stopY]);
    axis([startX stopX startY stopY])

    pause(0.01) % Pause here for 0.01 seconds
end
```

The included pause is used to make the cine loop play at a desired frame rate. It is also possible to run the cine loop at the original frame rate. A simple alteration of the code above would achieve this by displaying a frame only if the timing is correct.

```
figure; % Create figure
figureID = gca; % Retrieves ID for current axes
tic
for k=1:nFrames
    timePassed=toc;
    if timePassed<(timePerFrame*k+1) && timePassed>(timePerFrame*k-1)
        hilbertFrame = hilbert(rfData(:,:,k)); % HT on frame k
        amplitudeFrame = abs(hilbertFrame); % Extract the amplitude
        bmode = log10(amplitudeFrame); % Log compress
        imageHandle = imagesc(figureID,bmode); % Display image
        caxis(figureID,[1.5 4]); % Set colormap display limits
        colormap(gray) % Change colormap to grayscale
        % Change the x- and y-ranges for the displayed data
        set(imageHandle, 'XData', [startX, stopX]);
        set(imageHandle, 'YData', [startY, stopY]);
        axis([startX stopX startY stopY])

        drawnow; % Pause-command not needed but drawnow forces MATLAB to
        % display the image before continuing
    end
end
```

---

[2]This example works just fine for visualizing a series of images as a video, but it is generally advisable to avoid using for-loops for this purpose. MATLAB offers native support to write a series of images to an AVI file using the function VideoWriter. This may be a more suitable approach in some circumstances.

20.5 CONCLUSION

In this case study we have demonstrated how to read ultrasound raw data from an arbitrary, but predefined, file format. Different manufacturers all have their own binary file formats when exporting ultrasound RF data. Therefore, the knowledge on how to write scripts similar to the ones used above provides a flexible approach in accessing these types of data. This case study also provided an example of how to visualize an ultrasound B-Mode image, reconstructed from RF data. If the user needs to interact with the images, which is common, we suggest combining this knowledge with the use of a graphical user interface (GUI) as this significantly facilitates the work flow.

MATLAB toolboxes used in this chapter:
Signal Processing Toolbox

Index of the in-built MATLAB functions used:

abs	fopen	log10	strfind
char	fread	pause	tic
drawnow	hilbert	reshape	toc
fclose	imagesc	split	xmlread
find			

Bibliography

[1] AAPM. MEDPHYS 3.0—Physics for Every Patient. `https://w3.aapm.org/medphys30/index.php`.

[2] E. Samei, T. Pawlicki, D. Bourland, et al. Redefining and reinvigorating the role of physics in clinical medicine: a report from the AAPM Medical Physics 3.0 Ad Hoc Committee. *Med. Phy.*, 45(9):e783–e789, 2018.

[3] P. Nowik, R. Bujila, G. Poludniowski, and A. Fransson. Quality control of CT systems by automated monitoring of key performance indicators: a two-year study. *J. Appl. Clin. Med. Phys.*, 16(4):254–265, 2015.

[4] ESR. AI blog. `https://ai.myesr.org/`.

[5] Digital Imaging and Communication in Medicine (DICOM). The DICOM Standard, NEMA, 2020. `https://www.dicomstandard.org/current/`.

[6] Health Level Seven International (HL7). HL7, 2020. `https://www.hl7.org/`.

[7] (IHE) Integrating the Healthcare Enterprise. IHE, 2020. `https://wiki.ihe.net/index.php/Radiation_Exposure_Monitoring`.

[8] (IAEA) The International Atomic Energy Agency. Diagnostic Reference Levels (DRL's). `https://www.iaea.org/resources/rpop/health-professionals/radiology/diagnostic-reference-levels`.

[9] J.M. Boone, K.J. Strauss, D.D. Cody, C.H. McCollough, M.F. McNitt-Gray, and T.L. Toth. Size-specific dose estimates (SSDE) in pediatric and adult body CT examination. American Association of Physicists in Medicine, 2011.

[10] C.H. McCollough, M.B. Donovan, M.Bostani, S.Brady, K.Boedeker, and J.M. Boone, et al. Use of water equivalent diameter for calculating patient size and size-specific dose estimates (SSDE) in CT. Technical Report, College Park, MD: American Association of Physicists in Medicine, 2014.

[11] Dicom Port AB. autoQA™, 2020. `https://www.dicom-port.com/products/autoqa/`.

[12] Siemens Healthineers. Siemens teamplay. `https://www.siemens-healthineers.com/digital-health-solutions/digital-solutions-overview/service-line-managment-solutions/teamplay`.

[13] Siemens Healthineers. Siemens Guardian. `https://www.siemens-healthineers.com/services/customer-services/asset-evolution-services/siemens-guardian-program`.

[14] J. Andersson, W. Pavlicek, R. Al-Senan, et al. Estimating patient organ dose with computed tomography: a review of present methodology and required DICOM information. Technical Report 246, American Association of Physicsts in Medicine, 2019.

[15] International Electrotechnical Commission, Geneva. *IEC 61267:2005 Medical diagnostic X-ray equipment—Radiation conditions for use in the determination of characteristics.*

[16] IPEM Report 32 Part VII—Measurement of the Performance Characteristics of Diagnostic X-Ray Systems: Digital Imaging Systems. Technical Report, Institute of Physics and Engineering in Medicine, 2010.

[17] International Electrotechnical Commission, Geneva. *IEC 62220-1-1:2015 Medical electrical equipment—Characteristics of digital x-ray imaging devices—Part 1-1: Determination of the detective quantum efficiency—Detectors used in radiographic imaging.*

[18] D. Kennedy, D.R Hipp, and J. Mistachkin. SQLite. `https://www.sqlite.org/index.html`.

[19] R. Bujila, A. Omar, and G. Poludniowski. A validation of spekpy: a software toolkit for modelling x-ray tube spectra. *Phys. Med.*, 75:44–54, 2020.

[20] G.J. Salomons and D. Kelly. A survey of canadian medical physicists: software quality assurance of in-house software. *J. Appl. Clin. Med. Phys.*, 16(1):336–348, January 2015.

[21] H. Jones and M. Herron. The use of mobile apps as a form of healthcare in the NHS. `https://www.bristows.com/news/the-use-of-mobile-apps-and-software-as-a-form-of-healthcare-within-the-nhs/`.

[22] J. Nixon. The changing landscape for medical devices. `https://www.brownejacobson.com/training-and-resources/resources/legal-updates/2017/06/the-changing-regulatory-landscape-for-medical-devices`. June, 2017.

[23] European Commission. New regulations. `https://ec.europa.eu/growth/sectors/medical-devices/new-regulations\_en`.

[24] B.V. Vooren. EU medical devices regulation series: interpreting the industrial scale concept. `https://www.insidemedicaldevices.com/2017/01/eu-medical-devices-regulation-series-interpreting-the-industrial-scale-concept/`.

[25] MHRA. Health institution exemption for IVDR/MDR. `https://www.gov.uk/government/consultations/health-institution-exemption-for-ivdrmdr`.

[26] J.P. McCarthy. MDR—The health institution exemption and MHRA draft guidelines. `https://www.ipem.ac.uk/Portals/0/PDF/SCOPE_September2018_WEB.pdf`, 2018.

[27] J.P. McCarthy and A. Davie. The new medical devices regulation: interview with Justin McCarthy. `https://www.ipem.ac.uk/Portals/0/Documents/Publications/SCOPE/SCOPE_SEPT2017_LR.pdf`, 2017.

[28] European Commission. Commission communication in the framework of the implementation of the Council Directive 93/42/EEC concerning medical devices (2017/C 389/03). `https://eur-lex.europa.eu/legal-content/EN/TXT/?uri=uriserv:OJ.C_.2017.389.01.0029.01.ENG&toc=OJ:C:2017:389:TOC`.

[29] D. Willis, A. Green, P. Cosgriff, P. Ganney, and R. Trouncer. In-house development of medical software. `https://www.ipem.ac.uk/Portals/0/Documents/Publications/SCOPE/SCOPE_MAR2016_LR.pdf`, 2016.

[30] FDA. Reclassification (of medical devices). `https://www.fda.gov/MedicalDevices/DeviceRegulationandGuidance/Overview/ClassifyYourDevice/ucm080412.htm`.

[31] FDA. Premarket Approval (PMA). `https://www.fda.gov/MedicalDevices/DeviceRegulationandGuidance/HowtoMarketYourDevice/PremarketSubmissions/PremarketApprovalPMA/default.htm`.

[32] R. Cepeda. Mobile Medical Apps & FDA Rules: What Is the Latest in 2017? `https://www.fda.gov/MedicalDevices/DeviceRegulationandGuidance/HowtoMarketYourDevice/PremarketSubmissions/PremarketApprovalPMA/default.htm`.

[33] Z. Brennan. FDA's Regulation of CDS Software: Will Physicians Have to Understand the Underlying Algorithms? `https://www.raps.org/news-and-articles/news-articles/2018/4/fdas-regulation-of-cds-software-will-physicians`.

[34] A. Mulero. Industry Raises Concerns with FDA Draft Guidance on Clinical Decision Support Software. `https://www.raps.org/news-and-articles/news-articles/2018/3/industry-raises-concerns-with-fda-draft-guidance-o`.

[35] FDA. A-Z List of Regulated Products & Procedures. `https://www.fda.gov/Radiation-EmittingProducts/RadiationEmittingProductsandProcedures/ucm135878.htm`.

[36] FDA. Guidance for the Submission of Premarket Notifications for Emission Computed Tomography Devices and Accessories (SPECT and PET) and Nuclear Tomography Systems. `https://www.fda.gov/medicaldevices/deviceregulationandguidance/guidancedocuments/ucm073794.htm`.

[37] FDA. Guidance for the Submission of Premarket Notifications for Medical Image Management Devices—Guidance for Industry. `https://www.fda.gov/regulatory-information/search-fda-guidance-documents/guidance-submission-premarket-notifications-medical-image-management-devices-guidance-industry`.

[38] Imaging Technology News. FDA Proposes to Reclassify Some Types of Radiology Image Software Analyzers. `https://www.itnonline.com/content/fda-proposes-reclassify-some-types-radiology-image-software-analyzers`.

[39] FDA. Device establishment registration exemptions regarding 21cfr807.65. `https://www.ecfr.gov/cgi-bin/text-idx?SID=8acc2e713090fcafb6c7d39de36e4418&mc=true&node=pt21.8.807&rgn=div5#sp21.8.807.d`.

[40] J.H. Sinard and P. Gershkovich. Custom software development for use in a clinical laboratory. *J. Pathol. Inform.*, 3(1):44, 2012.

[41] J.J. Smith. Medical Imaging: The Basics of FDA Regulation. `https://www.mddionline.com/medical-imaging-basics-fda-regulation`.

[42] FDA. Examples of Pre-Market Submissions that Include MMAs Cleared or Approved by FDA. `https://www.fda.gov/MedicalDevices/DigitalHealth/MobileMedicalApplications/ucm368784.htm`.

[43] W. Imam. Differences and similarities between FDA 21 CFR Part 820 and ISO 13485. `https://advisera.com/13485academy/blog/2017/10/05/differences-and-similarities-between-fda-21-cfr-part-820-and-iso-13485/`.

[44] M. Trevino. Medical Device Single Audit Program (MDSAP)—What Manufacturers Need To Know. https://www.meddeviceonline.com/doc/understanding-the-medical-device-single-audit-program-mdsap-0001.

[45] FDA. Off-The-Shelf Software Use in Medical Devices. https://www.fda.gov/media/71794/download.

[46] C. Michaud. Got SOUP?—Part 2—OS, Drivers, Runtimes. https://blog.cm-dm.com/post/2013/05/24/Got-SOUP-Part-2-OS\%2C-Drivers\%2C-Runtimes.

[47] Johner Institute. SOUP—Software of Unknown Provenance. https://www.johner-institute.com/articles/software-iec-62304/soup-and-ots/.

[48] C. Hobbs. Clear SOUP and COTS Software for Medical Device Development. https://blackberry.qnx.com/content/dam/qnx/whitepapers/2011/qnx_med_cots_soup.pdf.

[49] MathWorks. FDA software validation. https://uk.mathworks.com/solutions/medical-devices/fda-software-validation.html, 2017.

[50] The Business Dictionary. What is 'Best practice'. http://www.businessdictionary.com/definition/best-practice.html.

[51] European Commission. Medical devices guidance. https://ec.europa.eu/growth/sectors/medical-devices/guidance_en.

[52] MHRA. Regulatory guidance for medical devices. https://www.gov.uk/government/collections/regulatory-guidance-for-medical-devices.

[53] Swedish Medical Products Agency. Medical information systems—guidance for qualification and classification of standalone software with a medical purpose. https://lakemedelsverket.se/upload/eng-mpa-se/vagledningar_eng/medical-information-system-guideline.pdf.

[54] FDA. The essential list of guidances for medical devices. https://blog.cm-dm.com/pages/The-essential-list-of-guidances-for-software-medical-devices#FDA.

[55] IMDRF. Software as a Medical Device (SaMD): Application of Quality Management System. http://www.imdrf.org/docs/imdrf/final/technical/imdrf-tech-151002-samd-qms.pdf.

[56] FDA. General Principles of Software Validation: Guidance for Industry and FDA Staff. https://www.fda.gov/regulatory-information/search-fda-guidance-documents/general-principles-software-validation.

[57] B. Wichmann, G. Parkin, and R. Barker. Software Support for Metrology Best Practice Guide No. 1: Validation of Software in Measurement Systems. http://publications.npl.co.uk/npl_web/pdf/dem_es14.pdf.

[58] Wikipedia. Best coding practices. https://en.wikipedia.org/wiki/Best_coding_practices.

[59] WikiBooks. MATLAB Programming. https://en.wikibooks.org/wiki/MATLAB_Programming.

[60] R. Johnson. MATLAB Style Guidelines 2.0. `https://se.mathworks.com/matlabcentral/fileexchange/46056-matlab-style-guidelines-2-0`.

[61] J. McCarthy. Medical Device and Health Software: Standards and regulations now and in the future. `http://projects.npl.co.uk/metromrt/news-events/20150420-21_workshop/12-mccarthy.pdf`.

[62] R.A. Miller. Summary recommendations for responsible monitoring and regulation of clinical software systems. *Ann. Intern. Med.*, 127(9):842, November 1997.

[63] E. Hoxey. Updated guidance on medical device borderline and classification issued. `http://compliancenavigator.bsigroup.com/en/medicaldeviceblog/updated-guidance-on-medical-device-borderline-and-classification-issued/`.

[64] EDPB. European Data Protection Board. `https://edpb.europa.eu/`.

[65] ICO. Information Commissioner's Office. `https://ico.org.uk/for-organisations/guide-to-data-protection/guide-to-the-general-data-protection-regulation-gdpr/`.

[66] HIMSS. Healthcare Information and Management Systems Society. `https://www.himss.org/library/interoperability-standards/security-standards`.

[67] European Commission. Guidelines on Data Protection Impact Assessment (DPIA) (wp248rev.01). `https://ec.europa.eu/newsroom/article29/item-detail.cfm?item_id=611236`.

[68] Information Commissioner's Office. Anonymisation—code of practice. `https://ico.org.uk/media/1061/anonymisation-code.pdf`.

[69] Finnish Social Science Data Archive. Anonymisation and Personal Data. `https://www.fsd.tuni.fi/aineistonhallinta/en/anonymisation-and-identifiers.html#bases-of-anonymisation`.

[70] Analytic Imaging Diagnostics Arena. AIDA data sharing policy. `https://datasets.aida.medtech4health.se/sharing/`.

[71] Food and Drug Administration (FDA). *General principles of software validation; final guidance for industry and FDA staff.* US Department of Health and Human Services, 2002.

[72] International Electrotechnical Commission (IEC). *Medical device software—Software life cycle processes, International Standard IEC 62304:2006AMD1:2015.* IEC, Geneva, 2015.

[73] Therapeutic Goods Administration (TGA). *Consultation: Regulation of software, including Software as a Medical Device (SaMD).* Australian Government, Department of Health, 2019.

[74] M.S. Huq, B. Fraass, P. Dunscombe, et al. Application of risk analysis methods to radiation therapy quality management: report of AAPM task group 100. *Med. Phys.*, 43(7):4209–4262, 2016.

[75] H. Delis, K. Christaki, B. Healy, et al. Moving beyond quality control in diagnostic radiology and the role of the clinically qualified medical physicist. *Physica Medica*, 41:104–108, 2017.

[76] I.J. Das. *Radiochromic Film: Role and Applications in Radiation Dosimetry*. CRC Press, Florida, 2017.

[77] T. Kairn, N. Hardcastle, J. Kenny, et al. EBT2 radiochromic film for quality assurance of complex IMRT treatments of the prostate: micro-collimated IMRT, RapidArc, and TomoTherapy. *Australas. Phys. Eng. Sci. Med.*, 34(3):333, 2011.

[78] T. Kairn, D. Papworth, S.B. Crowe, et al. Dosimetric quality, accuracy, and deliverability of modulated radiotherapy treatments for spinal metastases. *Med. Dosim.*, 41(3):258–266, 2016.

[79] J.E. Morales, R. Hill, S.B. Crowe, et al. A comparison of surface doses for very small field size x-ray beams: Monte Carlo calculations and radiochromic film measurements. *Australas. Phys. Eng. Sci. Med.*, 37(2):303–309, 2014.

[80] N.D. Middlebrook, B. Sutherland, and T. Kairn. Optimization of the dosimetric leaf gap for use in planning VMAT treatments of spine SABR cases. *J. Appl. Clin. Med. Phys.*, 18(4):133–139, 2017.

[81] T. Geber, M. Gunnarsson, and S. Mattsson. Eye lens dosimetry for interventional procedures–relation between the absorbed dose to the lens and dose at measurement positions. *Radiat. Meas.*, 46(11):1248–1251, 2011.

[82] R.Y.L. Chu, G. Thomas, and F. Maqbool. Skin entrance radiation dose in an interventional radiology procedure. *Health Phys.*, 91(1):41–46, 2006.

[83] A. Karambatsakidou, A. Omar, B. Chehrazi, et al. Skin dose, effective dose and related risk in transcatheter aortic valve implantation (TAVI) procedures: is the cancer risk acceptable for younger patients? *Radiat. Prot. Dosim.*, 169(1-4):225–231, 2016.

[84] E.R. Giles and P.H. Murphy. Measuring skin dose with radiochromic dosimetry film in the cardiac catheterization laboratory. *Health phys.*, 82(6):875–880, 2002.

[85] M.L. Kirkwood, G.M. Arbique, J.B. Guild, et al. Radiation-induced skin injury after complex endovascular procedures. *J. Vasc. Surg.*, 60(3):742–748, 2014.

[86] S.C. Peet, R. Wilks, T. Kairn, et al. Calibrating radiochromic film in beams of uncertain quality. *Med. phys.*, 43(10):5647–5652, 2016.

[87] O. Rampado, E. Garelli, S. Deagostini, and R. Ropolo. Dose and energy dependence of response of Gafchromic XR-QA film for kilovoltage x-ray beams. *Phys. Med. Biol.*, 51(11):2871, 2006.

[88] S. Devic, N. Tomic, and D. Lewis. Reference radiochromic film dosimetry: review of technical aspects. *Physica Medica*, 32(4):541–556, 2016.

[89] B.P. McCabe, M.A. Speidel, T.L. Pike, et al. Calibration of Gafchromic XR-RV3 radiochromic film for skin dose measurement using standardized x-ray spectra and a commercial flatbed scanner. *Med. Phys.*, 38(4):1919–1930, 2011.

[90] M. Tamponi, R. Bona, A. Poggiu, and P. Marini. A new form of the calibration curve in radiochromic dosimetry. Properties and results. *Med. Phys.*, 43(7):4435–4446, 2016.

[91] J.H. Sinard and P. Gerhkovich. Custom software development for use in a clinical laboratory. *J. Pathol. Inform.*, 3:44, 2012.

[92] T. Aland, T. Kairn, and J. Kenny. Evaluation of a Gafchromic EBT2 film dosimetry system for radiotherapy quality assurance. *Australas. Phys. Eng. Sci. Med.*, 34(2):251–260, 2011.

[93] IEC. 61223-3-5: Evaluation and routine testing in medical imaging departments: Acceptance tests-imaging performance of computed tomography x-ray equipment. 2004.

[94] IEC. 61223-2-6: Evaluation and routine testing in medical imaging departments: Constancy tests-Imaging performance of computed tomography x-ray equipment. 2006.

[95] IEC. 60601-2-44, Ed 3.1: Medical electrical equipment: Particular requirements for the basic safety and essential performance of x-ray equipment for computed tomography. 2003.

[96] R.A. Brooks and G. Di Chiro. Statistical limitations in x-ray reconstructive tomography. *Med. phys.*, 3(4):237–240, 1976.

[97] D. Merzan, P. Nowik, G. Poludniowski, and R. Bujila. Evaluating the impact of scan settings on automatic tube current modulation in CT using a novel phantom. *Brit. J. Radiol.*, 90(1069):20160308, 2017.

[98] DICOM. *Supplement 94: Diagnostic X-ray Radiation Dose Reporting (Dose SR)*. Rosslyn, VA: National Electrical Manufacturers Association (NEMA), 2005.

[99] Digital Imaging and Communication in Medicine (DICOM). PS3.3 2019c—Information Object Definitions, NEMA, 2019.

[100] Digital Imaging and Communication in Medicine (DICOM). PS3.16 2019c—Content Mapping Resource, NEMA, 2019.

[101] E. Vañó, D.L. Miller, C.J. Martin, M.M. Rehani, et al. ICRP Publication 135: Diagnostic Reference Levels in Medical Imaging. *Annals of the ICRP*, 46(1):1–144, 2017.

[102] Swedish Radiation Safety Authority. Vägledning med bakgrund och motiv till SSMFS 2018:5 Strålsäkerhetsmyndighetens föreskrifter och allmänna råd om medicinska exponeringar.

[103] M.F. McNitt-Gray. AAPM/RSNA physics tutorial for residents: topics in CT. radiation dose in CT. *Radiographics*, 22(6):1541–1553, 2002.

[104] J.M. Boone. Reply to "Comment on the 'Report of AAPM TG 204: Size-specific dose estimates (SSDE) in pediatric and adult body CT examinations"'[AAPM Report 204, 2011]. *Med. phys.*, 39(7), Part 2:4615–4616, 2012.

[105] C.S. Burton, A. Malkus, F. Ranallo, and T.P. Szczykutowicz. Model-based magnification/minification correction of patient size surrogates extracted from CT localizers. *Med. phys.*, 2018.

[106] C.S. Burton and T.P. Szczykutowicz. Evaluation of aapm reports 204 and 220: Estimation of effective diameter, water-equivalent diameter, and ellipticity ratios for chest, abdomen, pelvis, and head CT scans. *J. Appl. Clin. Med. Phys.*, 19(1):228–238, 2018.

[107] O. Christianson, X. Li, D. Frush, and E. Samei. Automated size-specific CT dose monitoring program: Assessing variability in CT dose. *Med. phys.*, 39(11):7131–7139, 2012.

[108] C. Anam, F. Haryanto, R. Widita, I. Arif, G. Dougherty, and D. McLean. The impact of patient table on size-specific dose estimate (SSDE). *Australas. Phys Eng Sci Med*, 40(1):153–158, 2017.

[109] ICRP. *Statement on Tissue Reactions / Early and Late Effects of Radiation in Normal Tissues and Organs – Threshold Doses for Tissue Reactions in a Radiation Protection Context.* ICRP Publication 118, *Annals of the ICRP*, International Commission on Radiological Protection, 2012.

[110] G. Alm-Carlsson, D.R. Dance, L. DeWerd, et al. *Dosimetry in Diagnostic Radiology: An International Code of Practice.* IAEA Technical Reports Series no. 457. International Atomic Energy Agency, Vienna, AUT, 2007.

[111] NEMA PS3 / ISO 12052. *Digital Imaging and Communications in Medicine (DICOM) Standard,.* National Electrical Manufacturers Association, Rosslyn, VA, USA. (available at http://medical.nema.org/).

[112] IEC. Medical electrical equipment – Part 2-43: Particular requirements for the safety of x-ray equipment for interventional procedures. *International Electrotechnical Commission 60601, 2nd Ed.*, 2010.

[113] A. Omar, R. Bujila, A. Fransson, P. Andreo, and G. Poludniowski. A framework for organ dose estimation in x-ray angiography and interventional radiology based on dose-related data in DICOM structured reports. *Phys. Med. Biol.*, 61(8):3063–3083, Apr 2016.

[114] A. Omar, H. Benmakhlouf, M. Marteinsdottir, R. Bujila, P. Nowik, and P. Andreo. Monte Carlo investigation of backscatter factors for skin dose determination in interventional neuroradiology procedures. In *Medical Imaging 2014: Physics of Medical Imaging*, vol. 9033. (Bellingham: International Society for Optics and Photonics), 2014.

[115] A. Servomaa and M. Tapiovaara. Organ dose calculation in medical x ray examinations by the program PCXMC. *Radiat. Prot. Dosim.*, 80(1-3):213–219, 1998.

[116] M. Cristy and K.F. Eckerman. *Specific absorbed fractions of energy at various ages from internal photon sources. I. Methods.* Report/TM-8381/V1. Oak Ridge National Laboratory, Oak Ridge, TN, 1987.

[117] ICRU. Appendix F: PCXMC: a PC-based Monte Carlo program for calculating organ doses for patients in medical x-ray examinations. *J. ICRU.*, 5(2):100–102, 2005.

[118] J. Ferlay, M. Colombet, I. Soerjomataram, et al. Cancer incidence and mortality patterns in europe: estimates for 40 countries and 25 major cancers in 2018. *Eur. J. Cancer*, 2018.

[119] S. Ciatto, N. Houssami, D. Bernardi, et al. Integration of 3D digital mammography with tomosynthesis for population breast-cancer screening (storm): a prospective comparison study. *Lancet. oncol.*, 14(7):583–589, 2013.

[120] P. Skaane, A.I. Bandos, R. Gullien, et al. Comparison of digital mammography alone and digital mammography plus tomosynthesis in a population-based screening program. *Radiology,.* 267(1):47–56, 2013.

[121] S. Zackrisson, K. Lang, A. Rosso, et al. One-view breast tomosynthesis versus two-view mammography in the malmö breast tomosynthesis screening trial (mbtst): a prospective, population-based, diagnostic accuracy study. *Lancet. Oncol.*, 19(11):1493–1503, 2018.

[122] C.J. D'Orsi. *ACR BI-RADS Atlas: Breast Imaging Reporting and Data System.* American College of Radiology, 2013.

[123] D.D. Pokrajac, A.D.A. Maidment, and P.R. Bakic. Optimized generation of high resolution breast anthropomorphic software phantoms. *Med. phy.*, 39(4):2290–2302, 2012.

[124] L. Chen, J.M. Boone, A. Nosratieh, and C.K. Abbey. Nps comparison of anatomical noise characteristics in mammography, tomosynthesis, and breast CT images using power law metrics. In *Medical Imaging 2011: Physics of Medical Imaging*, vol. 7961, p. 79610F. International Society for Optics and Photonics, 2011.

[125] K. Perlin. An image synthesizer. *ACM Siggraph Computer Graphics*, 19(3):287–296, 1985.

[126] S. Worley. A cellular texture basis function. In *Proceedings of the 23rd annual conference on Computer graphics and interactive techniques*, pp. 291–294. ACM, 1996.

[127] M. Dustler, P. Bakic, H. Petersson, P. Timberg, A. Tingberg, and S. Zackrisson. Application of the fractal Perlin noise algorithm for the generation of simulated breast tissue. In *Medical Imaging 2015: Physics of Medical Imaging*, vol. 9412, p. 94123E. International Society for Optics and Photonics, 2015.

[128] M. Dustler, H. Förnvik, and K. Lång. Binary implementation of fractal perlin noise to simulate fibroglandular breast tissue. In *Medical Imaging 2018: Physics of Medical Imaging*, vol. 10573, p. 1057357. International Society for Optics and Photonics, 2018.

[129] T.F. Nano and I.A. Cunningham. xrTk: A MATLAB toolkit for x-ray physics calculations in medical imaging, available under GPL license from a git repository at. `http://dqe.robarts.ca/public`, port 443, 2019.

[130] Free Software Foundation, Inc. https://opensource.org/licenses/gpl-3.0.html, 2018. Accessed 2018-11-12.

[131] J.H. Siewerdsen, A.M. Waese, D.J. Moseley, S. Richard, and D.A. Jaffray. Spektr: a computational tool for x-ray spectral analysis and imaging system optimization. *Med. Phys.*, 31(11):3057 3067, 2004.

[132] G. Poludniowski, G. Landry, F. DeBlois, P.M. Evans, and F. Verhaegen. SpekCalc: a program to calculate photon spectra from tungsten anode x-ray tubes. *Phys. Med. Biol.*, 54:N433–N438, October 2009.

[133] M.J. Berger, J.H. Hubbell, S.M. Seltzer, et al. XCOM: Photon cross section database (version 1.3). `http://physics.nist.gov/xcom`, 2005.

[134] J.S. Hubbell and S.M. Seltzer. Tables of x-ray mass attenuation coefficients and mass energy-absorption coefficients from 1keV to 20 MeV for elements $Z = 1$ to 92 and 48 additional substances of dosimetric interest. *Radiat. Res.*, 136:147, 1993 (http://www.nist.gov/pml/data/xraycoef/).

[135] D.M. Tucker, G.T. Barnes, and D.P. Chakraborty. Semiempirical model for generating tungsten target x-ray spectra. *Med. Phys.*, 18(2):211–218, 1991.

[136] D.M. Tucker, G.T. Barnes, and X.Z. Wu. Molybdenum target x-ray spectra: a semiempirical model. *Med. Phys.*, 18:402–407, 1991.

[137] Medical electrical equipment—characteristics of digital x-ray imaging devimaging Part 1-1: Determination of the detective quantum efficiency—Detectors used in radiographic imaging. Medical Electrical Equipment IEC 62220-1-1, International Electrotechnical Commission, 2015.

[138] Medical electrical equipment—characteristics of digital x-ray imaging devices—Part 1-2: Determination of the detective quantum efficiency—Detectors used in mammography. Medical electrical equipment IEC 62220-1-2, International Electrotechnical Commission, 2007.

[139] H.E. Johns and J.R. Cunningham. *The Physics of Radiology, 4th Ed.* C.C. Thomas, Springfield, IL, 1983.

[140] F.H. Attix. *Introduction to Radiological Physics and Radiation Dosimetry.* John Wiley & Sons, New York, 1986.

[141] M.J. Yaffe and J.A. Rowlands. X-ray detectors for digital radiography. *Phys. Med. Biol.*, 42(1):1–39, 1997.

[142] ICRU 44. International Commission on Radiation Units and Measurements Report 44: Tissue substitutes in radiation dosimetry and measurement. 1989.

[143] A. Rose. *Vision: Human and Electronic.* Plenum Press, New York, 1974.

[144] I.A. Cunningham and R. Shaw. Signal-to-noise optimization of medical imaging systems. *Opt. Soc. Am. A., Optics and Image Science*, 16(3):621–632, 1999.

[145] I.A. Cunningham. *Handbook of Medical Imaging, Volume 1. Physics and Psychophysics, Chapter 2: Applied linear-systems theory.* SPIE Press, Washington, 2000.

[146] R.K. Swank. Absorption and noise in x-ray phosphors. *J. Appl. Phys.*, 44(9):4199–4203, 1973.

[147] C.S. Burton, J.R. Mayo, and I.A. Cunningham. Energy subtraction angiography is comparable to digital subtraction angiography in terms of iodine Rose SNR. *Med. Phys.*, 43(11):5925–5933, 2016.

[148] J.E. Gray, B.R. Archer, P.F. Butler, et al. Reference values for diagnostic radiology: application and impact. *Radiology*, 235(2):354–358, 2005.

[149] W. Zhao, G. Ristic, and J.A. Rowlands. X-ray imaging performance of structured cesium iodide scintillators. *Med. Phys.*, 31(9):2594–2605, 2004.

[150] C. Michail. Image quality assessment of a CMOS/Gd^2O^2S:Pr,Ce,F x-ray sensor. *J. Sens.*, 2015.

[151] J.S. Hubbell. Compilation of photon cross-sections: some historical remarks and current status. *X-Ray Spectrom.*, 28(4):215–223, 1999.

[152] American College of Radiology. Acceptance testing and quality assurance procedures for magnetic resonance imaging facilities. `https://www.aapm.org/pubs/reports/RPT_100.pdf`. American Association of Physicists in Medicine, Report no. 100.

[153] G.D. Clarke. Overview of the ACR MRI accreditiation phantom. `https://www.aapm.org/meetings/99AM/pdf/2728-58500.pdf`. University of Texas Southwestern Medical Center at Dallas.

[154] American College of Radiology. Phantom test guidance for use of the large MRI phantom for the ACR MRI accreditation program. `https://www.acraccreditation.org/-/media/ACRAccreditation/Documents/MRI/LargePhantomGuidance.pdf?la=en`.

[155] A. Krizhevsky, I. Sutskever, and G.E. Hinton. Imagenet classification with deep convolutional neural networks. In F. Pereira, C.J.C. Burges, L. Bottou, and K.Q. Weinberger, editors, *Advances in Neural Information Processing Systems 25*, pp. 1097–1105. Curran Associates, Inc., 2012.

[156] M. Chillarón, V. Vidal, D. Segrelles, I. Blanquer, and G. Verdú. Combining grid computing and docker containers for the study and parametrization of CT image reconstruction methods. *Procedia Computer Science*, 108:1195 – 1204, 2017. International Conference on Computational Science, ICCS 2017, 12-14 June 2017, Zurich, Switzerland.

[157] E.A. Eisenhauer, P. Therasse, J. Bogaerts, et al. New response evaluation criteria in solid tumours: Revised recist guideline (version 1.1). *Eur. J. Cancer*, 45(2):228–247, 1 2009.

[158] J. Wu, Y. Cui, X. Sun, et al. Unsupervised clustering of quantitative image phenotypes reveals breast cancer subtypes with distinct prognoses and molecular pathways. *Clin. Cancer Res.*, 23(13):3334–3342, 2017.

[159] D.S. Marcus, T.R. Olsen, M. Ramaratnam, and R.L. Buckner. The extensible neuroimaging archive toolkit—an informatics platform for managing, exploring, and sharing neuroimaging data. *Neuroinformatics*, 5(1):11–33, 2007.

[160] W.W. Nichols and M.F. O'Rourke. *McDonald's Blood Flow in Arteries*. Edward Arnold, London, 2005.

[161] M. Cinthio, Å.R. Ahlgren, T. Jansson, A. Eriksson, H.W. Persson, and K. Lindström. Evaluation of an ultrasonic echo-tracking method for measurements of arterial wall movements in two dimensions. *IEEE Trans. Ultrason. Ferroelectr. Freq. Control*, 52(8):1300–1311, 2005.

[162] M. Cinthio, Å.R. Ahlgren, J. Bergkvist, T. Jansson, H.W. Persson, and Lindström K. Longitudinal movements and resulting shear strain of the arterial wall. *Am. J. Physiol. - Heart Circ. Physiol.*, 291(1):H394–H402, 2006.

[163] Å.R. Ahlgren, M. Cinthio, S. Steen, et al. Longitudinal displacement and intramural shear strain of the porcine carotid artery undergo profound changes in response to catecholamines. *Am. J. Physiol. - Heart Circ. Physiol.*, 302(5):H1102–H1115, 2012.

[164] S. Svedlund and L.M. Gan Longitudinal common carotid artery wall motion is associated with plaque burden in man and mouse. *Atherosclerosis*, 217(1):120–124, 2011.

[165] H.S. Taivainen, H. Yli-Ollila, M. Juonala, et al. Influence of cardiovascular risk factors on longitudinal motion of the common carotid artery wall. *Atherosclerosis*, 272(1):54–59, 2018.

[166] G. Zahnd, L.J. Maple-Brown, K. O'Dea, et al. Longitudinal displacement of the carotid wall and cardiovascular risk factors: associations with aging, adiposity, blood pressure and periodontal disease independent of cross-sectional distensibility and intima-media thickness. *Ultrasound Med. Biol.*, 38(10):1705–1715, 2012.

[167] J. Albinsson, S. Brorsson, Å.R. Ahlgren, and Cinthio M. Improved tracking performance of lagrangian block-matching methodologies using block expansion in the time domain – in silico, phantom and in vivo evaluations using ultrasound images. *Ultrasound Med. Biol.*, 40(10):2508–2520, 2014.

[168] E.I. Cespedes, Y. Huang, J. Ophir, and S. Spratt. Methods for estimation of subsample time delays of digitized echo signals. *Ultrason. lmaging*, 17(2):142–171, 1995.

[169] B.J. Geiman, L.N. Bohs, M.E. Anderson, S.M. Breit, and G.E. Trahey. A novel interpolation strategy for estimating subsample speckle motion. *Phys. Med. Biol.*, 45(6):1541–1552, 2000.

[170] J. Albinsson, Å.R. Ahlgren, T. Jansson, and M. Cinthio. A combination of parabolic and grid slope interpolation for 2D tissue displacement estimations. *Med. Biol. Eng. Comput.*, 55(8):1327–1338, 2017.

[171] M. Cinthio and Å.R. Ahlgren. Intra-observer variability of longitudinal movement and intramural shear strain measurements of the arterial wall using ultrasound non-invasively in vivo. *Ultrasound Med. Biol.*, 36(5):697–704, 2010.

[172] B.H. Friemel, L.N. Bohs, and G.E. Trahey. Relative performance of two-dimensional speckle-tracking techniques: normalized correlation, non-normalized correlation and sum-absolute-difference. In *Proceedings of IEEE Ultrasonic Symposium*, pp. 1481–1484. IEEE, 1995.

[173] M. Cinthio, T. Jansson, Å.R. Ahlgren, H.W. Persson, and K. Lindström. New improved and modified method for measurements of arterial wall movements in longitudinal and radial directions. In *Proceedings in Biomedizinische Technik supp vol 1, Part II*, vol. 50, pp. 869–870, 2005.

[174] M. Cinthio, T. Jansson, Å.R. Ahlgren, K. Lindström, and H.W. Persson. A method for arterial diameter change measurements using ultrasonic b-mode data. *Ultrasound Med. Biol.*, 36(9):1504–1513, 2010.

[175] J. Kirkhorn. Introduction to IQ-demodulation of RF-data. `http://folk.ntnu.no/htorp/Undervisning/TTK10/IQdemodulation.pdf`.

[176] L. Marple. Computing the discrete-time "analytic" signal via FFT. *IEEE T. Signal Proces.*, 47(9):2600–2603, 1999.

Index

T - #0507 - 071024 - C292 - 254/178/13 - PB - 9780367642624 - Gloss Lamination